Stochastic Partial Differential Equations for Computer Vision with Uncertain Data

Synthesis Lectures on Visual Computing
Computer Graphics, Animation, Computational Photography, and Imaging

Editor
Brian A. Barsky, *University of California Berkeley*

This series presents lectures on research and development in visual computing for an audience of professional developers, researchers, and advanced students. Topics of interest include computational photography, animation, visualization, special effects, game design, image techniques, computational geometry, modeling, rendering, and others of interest to the visual computing system developer or researcher.

Stochastic Partial Differential Equations for Computer Vision with Uncertain Data
Tobias Preusser, Robert M. Kirby, and Torben Pätz

ISBN: 978-3-031-01466-6   paperback
ISBN: 978-3-031-02594-5   ebook

DOI: 10.1007/978-3-031-02594-5

A Publication in the Springer series
Synthesis Lectures on Visual Computing: Computer Graphics, Animation,
Computational Photography, and Imaging

Lecture #28
Series Editor: Brian A. Barsky, *University of California Berkeley*
Series ISSN
Print 2469-4215   Electronic 2469-4223

Stochastic Partial Differential Equations for Computer Vision with Uncertain Data

Tobias Preusser
Jacobs University Bremen and Fraunhofer MEVIS Bremen

Robert M. Kirby
University of Utah, Salt Lake City

Torben Pätz
Jacobs University Bremen and Fraunhofer MEVIS Bremen

SYNTHESIS LECTURES ON VISUAL COMPUTING: COMPUTER GRAPHICS, ANIMATION, COMPUTATIONAL PHOTOGRAPHY, AND IMAGING #28

ABSTRACT

In image processing and computer vision applications such as medical or scientific image data analysis, as well as in industrial scenarios, images are used as input measurement data. It is good scientific practice that proper measurements must be equipped with error and uncertainty estimates. For many applications, not only the measured values but also their errors and uncertainties, should be—and more and more frequently are—taken into account for further processing. This error and uncertainty propagation must be done for every processing step such that the final result comes with a reliable precision estimate.

The goal of this book is to introduce the reader to the recent advances from the field of uncertainty quantification and error propagation for computer vision, image processing, and image analysis that are based on partial differential equations (PDEs). It presents a concept with which error propagation and sensitivity analysis can be formulated with a set of basic operations. The approach discussed in this book has the potential for application in all areas of quantitative computer vision, image processing, and image analysis. In particular, it might help medical imaging finally become a scientific discipline that is characterized by the classical paradigms of observation, measurement, and error awareness.

This book is comprised of eight chapters. After an introduction to the goals of the book (Chapter 1), we present a brief review of PDEs and their numerical treatment (Chapter 2), PDE-based image processing (Chapter 3), and the numerics of stochastic PDEs (Chapter 4). We then proceed to define the concept of stochastic images (Chapter 5), describe how to accomplish image processing and computer vision with stochastic images (Chapter 6), and demonstrate the use of these principles for accomplishing sensitivity analysis (Chapter 7). Chapter 8 concludes the book and highlights new research topics for the future.

KEYWORDS

image processing, computer vision, stochastic images, uncertainty quantification, stochastic partial differential equation, polynomial chaos

Contents

Preface

Over the past two decades, the topic of uncertainty quantification within simulation science has emerged as a requirement for the advancement of many scientific and engineering endeavors. The confluence of our ability to sense and measure data at scale, advances in applied mathematics and data science, and the power of computing has enabled us to reexamine the simulation science pipeline in terms of errors and uncertainties.

This work started with our attempt to consider the world of image processing and computer vision in light of new mathematical perspectives and algorithms devised within the uncertainty quantification world. We soon realized that, in principle, it was possible to formulate error propagation and uncertainty quantification in computer vision and image processing as an integral part of basal operations. Working out the details for the respective image processing models and their numerical treatments has taken several years. This book summarizes our efforts since we initiated this field with a publication entitled *Building Blocks for Computer Vision with Stochastic Partial Differential Equations* almost a decade ago.

The target group for this book is researchers starting at an advanced graduate level who may have existing knowledge about, on the one hand, computer vision/image processing with PDEs or, on the other hand, the numerics of SPDEs. This book is meant to represent a toolbox for initiating research in this area. As a summary of our efforts, we have not written this book from the perspective of competing with other methods. Correspondingly, we do not show all the extensive connections with existing approaches for stochastic computer vision or to non-PDE-based computer vision approaches such as graph cuts, etc. Our examples have a "bottom-up" characteristic and not that of emphasizing the highly challenging and large-scale application problems that we acknowledge exist in this area. We will give only some anchor points to the stochastic image processing community related to Bayesian inference, etc. The reader should keep in mind that going deep into this is not our focus; our focus is to build momentum and interest in the area of stochastic PDEs in computer vision.

In terms of what is needed to read this book, we assume that all readers are familiar with linear algebra and calculus in several variables. For those readers without a background in numerical methods for PDEs and without a background in probability, we recommend reading through the book linearly and also following up on the background reading references given herein. Readers who are familiar with classical (deterministic) PDE-based computer vision and image processing may chose to skip Chapter 3. Those who are experienced in numerics of PDEs and/or numerics of stochastic PDEs may skip parts of Chapter 4. The central part of our approach to SPDE-based computer vision and image processing is discussed in Chapter 5. An easy entry point into the applications of stochastic PDEs to computer vision and image processing is the sensitivity analysis in Chapter 7.

Any work of this size and scope has benefitted from the involvement of many people both indirectly and directly. We wish to thank our collaborators that inspired us with discussions about computer vision and uncertainty quantification in various models and application areas. We also thank the various faculty, students, and colleagues at the SCI Institute (University of Utah), Fraunhofer MEVIS (Bremen, Germany), and Jacobs University (Bremen, Germany) with whom we sharpened our ideas. In addition, we would like to thank the various Federal Funding Agencies that have supported our research efforts over the years. The papers we reference that we coauthored detail those acknowledgments. Lastly, we would like to thank our spouses, without whose patience and encouragement, we would probably not have made it this far.

Tobias Preusser, Robert M. Kirby, and Torben Pätz
June 2017

Notation

CHAPTER 1

Introduction

For many decades, a scientific culture of measurements, modeling, and error handling has been developing in the classic sciences such as physics, chemistry, and astronomy. Nowadays, technological advances are rapidly improving measurement devices in these disciplines with the goal of higher energies, higher resolutions, and more detail. At the same time, the awareness of the limitations of measurements and the handling of errors remains the basis for the evaluation of hypotheses and judgments on the quality and validity of claims.

With the dawn of the digital age more than 50 years ago, digital measurements and, in particular, digital imaging and digital image processing have become important components of research in all the natural sciences. Almost all measurements taken today are acquired digitally. Together with the recent increases in computer power used for data evaluations and with mathematical modeling, modern digital measurement techniques have led to significant state-of-the-art advances in the natural sciences.

The fields of computer vision and image processing have emerged from the digitalization of measurements and observations as intermediates between image data acquisition and analysis or modeling. The central purpose of these fields is to support the analysis and evaluation of the measurement data, primarily to extract quantitative information, resulting in tasks such denoising, segmentation, registration, inpainting, etc. In the past decades, myriads of operators and tools for computer vision, image processing, and image analysis have been developed.

As part of the scientific culture, it has become good scientific practice to equip measurements of any kind with information about the measurement error, allowing one to estimate the reliability of the measurements. Although the uncertainty from reading analog scales is removed in digital measurements, they are still subject to electronic noise and other uncertainties intrinsic to the physical measurement process. In digital imaging, these physical measurements focus on, e.g., photon density, time-of-flight of the waves, spin, and absorption. Although measurement errors, noise, and the uncertainty in the data have been investigated for specific imaging devices, the step of quantifying the error or uncertainty after an image processing operator has been applied can be very difficult—in particular for nonlinear operators—and thus is often omitted. Omission of this step leads to a loss of information about the influence of the input error on the results of the image processing pipeline, results that are then used as input for further downstream steps of analysis or modeling.

Radiology is one scientific discipline that has changed fundamentally and benefited enormously from the digitization of imaging. It has been only slightly more than a century since radiology was established with the discovery of X-rays by W. C. Röntgen in 1895/1896. Despite the digitalization that has taken place, radiology is a science that is still very much centered around the human eye as the perceiving device. Also, in contrast to other scientific disciplines, the scientific culture of error awareness and error handling for the measuring processes of radiology and medicine seems to be largely missing.

This book summarizes recent advances in the field of uncertainty quantification (UQ) and error propagation as applied to computer vision, image processing, and image analysis—advances that are based on the use of partial differential equations (PDEs) and UQ in image processing. It presents a concept with which error propagation and sensitivity analysis can be formulated with a set of basic operations. This approach has the potential for application in all areas of quantitative computer vision, image processing, and image analysis. In particular, it might help medical imaging finally become a scientific discipline that is characterized by the classical paradigms of observation, measurement, and error awareness.

Computer Vision, Image Processing, and Image Analysis with Partial Differential Equations.   The use of PDEs has become one of the major pillars of the fields of computer vision, image processing, and image analysis. The principal usage of PDEs is for identifying the pixel/voxel values of an image with a function that is then deformed via a PDE to obtain a filtered version of the original image. These filtered images are the solution of the formulated differential equation. The design of such PDEs has led to evolution models in which a given image is treated as initial data for a parabolic initial and boundary value problem. The time parameter of the process represents the so-called scale, which leads the image through the evolution process from the original image data to filtered versions of the image. The approaches presented in the past are too numerous to be briefly summarized. They range from image denoising, restoration, segmentation, and registration, to flow extraction, with further connections to models and simulations of physical and chemical processes that take data from images as input. For an overview of PDE-based computer vision and image processing, we refer the reader to [6, 17, 92].

In many cases, the image filtering results from the minimization of energy functionals or objective functions on the space of image functions. PDEs then result as the Euler-Lagrange equations that characterize the necessary condition for minima of the energy functionals. In contrast, parabolic PDEs are gradient descents that minimize the functionals and in which the initial image is used as a starting point for the energy descent. The design of energy functionals and their numerical minimization constitutes the art of PDE-based image processing. An energy functional should evaluate how well a proposed filtered or processed version of the image matches the properties of the desired result. In general, additional constraints in the form of regularizers are needed to make the minimization of the energy well-posed. The discretization and numerical solution are done with the classical approaches of the spectral or Galerkin methods, including the finite element, Fourier, or Wavelet Galerkin methods, and also with the classical finite difference method. Discretizations yield solutions, or minimizers, with robust algorithms the performance of which fit into the framework that is provided by the application field. For nonconvex energy functionals and nonlinear PDEs in particular, this task can be challenging.

The advantage of PDE-based methods over discrete methods is that they are a priori based on a continuous formulation of images, which, combined with their lattice (regular grid) structure, makes them amenable to finite difference or finite element methods. Also, for PDE-based methods, it is possible to study the existence, uniqueness, and regularity of solutions using well-known and powerful results from functional analysis. Finally, PDE-based methods naturally extend to the context of stochastic PDE-based image processing and computer vision in the case of image data that is uncertain and contains measurement errors. These three topics are the subject of the present book.

The other major pillar of image processing and computer vision is derived from probabilistic and statistical methods. Image analysis is closely related to spatial statistics and statistical models for images. The most prominent among these are Markov Random Fields and Bayesian image models. The Bayesian approach allows for the inclusion of prior knowledge into the analysis, thereby regularizing the ill-posed problems in image processing. Among the concepts from statistical and probabilistic image analysis are marked point processes, which are more general than Markov random fields as they allow for modeling relations between objects (pixels/voxels) without introducing a field lattice. This feature makes it easy to adapt the marked points during the process, i.e., points can be added and removed, which is not possible for Markov random fields on a fixed lattice. Furthermore, change point methods [13] allow for detecting anomalies/shocks/jumps in distributions. Such features are frequently used for 1D signal processing [1]. For the higher-dimensional images, i.e., 2D and 3D signals, change points are generalized to change curves and change surfaces. Qui devotes an entire book [84] to the application of change curves/surfaces in image analysis. A general introduction and overview to this rich field is given, e.g., in [113] and [32].

In applications of computer vision models, the user adjusts the parameters of the method multiple times to get satisfying results. Often, such parameter tuning must be done for every single data set. Thus, a result, e.g., a segmentation, depends not only on the input image, the selected segmentation method, and potential measurement errors, but also on the particular choice of the parameters by the user. Furthermore, a segmentation result may not be reproducible among different users. For medical applications, when different users segment a tumor with their individual segmentation parameters in different scans of a patient, the influence of the user parameters on the segmentation result can become important. In such cases, it is difficult to decide whether the segmentation result is different due to a growth of the tumor or because different segmentation parameters have been used. Since the further treatment of a patient is based on the medical RECIST classification [25, 100] of the segmentation result, sensitive segmentation results can have significant consequences. Thus, information about the stability of the segmentation with respect to the parameters can be useful for coming to an informed decision.

Finding a good parameterization of PDE-based image processing operators is a serious problem in image processing [83, 118]. Many authors have worked on methods for estimating optimal parameterization [58, 83, 119] or heuristics for a good parameterization when the optimal one is unknown [58]. Nevertheless, most of the work dealing with parameter estimation does not address the sensitivity of the methods with respect to changes in the parameters. Thus, a small deviation from the optimal parameterization or—even worse—a small error in the computation of the parametrization may lead to significantly deviating results.

Uncertainty Quantification in PDEs.   In many areas of mathematical modeling, (partial) differential equations are used to describe physical processes. The parameters of these equations represent physical properties of the object under investigation. In addition, boundary values and/or initial conditions further characterize the solution of the PDEs. In the spirit of the previously described scientific culture of observation, measurements, and error awareness, it must be noted that these data carry with them errors or uncertainty whenever they result from measurements or other sources that are uncertain.

In the past decade, major advances have been made in taking such uncertainties into account, thus shaping the research field of UQ. Significant theoretical foundations for UQ were laid many decades ago, but the increase in computing power has made the application of such concepts to real-world problems more feasible and thus triggered further progress in aspects of numerical analysis and implementation.

One of the fundamental concepts of UQ in the realm of PDEs is the identification of quantities under investigation with random variables or random fields, i.e., random variables that are indexed by a spatial location. Using such random fields in PDEs yields stochastic PDEs (SPDEs) with solutions that are again random fields that encode the distribution of solutions to the original PDE if the data vary stochastically.

The classical procedure to study the dependence on stochastic data numerically is the Monte Carlo sampling approach, which is amazingly robust but at the price of very slow convergence. If assumptions on the smoothness of the underlying processes can be made, sampling on regular, sparse, hierarchical, and adaptive grids may yield the desired information with much less effort. Moreover, sampling approaches share the property of nonintrusiveness: any existing code can easily and without many modifications be used to sample the parameter spaces.

Using a Hilbert-space structure on the stochastic space, Galerkin discretizations of the SPDEs yield an alternative approach. In this case, the stochastic dimension of the random fields is represented, e.g., by the Wiener-Askey polynomial chaos. The coefficients (i.e., the modes) of the stochastic representation can then be used to calculate the moments of the distribution of the result. For the physical dimension, any of the classical approaches, such as the finite element method or finite difference method, can be used. The resulting discrete equations are sparse systems of very high dimension, which is known as the *curse of dimension*. Sophisticated solvers and sparse tensor representations are being investigated to deal with this issue; this subfield of computational linear algebra continues to be a very active field of research.

In contrast to the sampling approaches, the stochastic Galerkin method is intrusive, i.e., existing code for the deterministic equations needs to be broken up and reengineered. The benefit is, however, that with few modes, a reasonably good approximation to the stochastic moments of the result can be achieved, whereas with sampling, the same level of accuracy would require much more effort. In this regard, the intrusiveness and the curse of dimension are fully justified.

Uncertainty Quantification in Computer Vision, Image Processing, and Image Analysis. Error propagation for PDE-based image processing, as described above, is feasible if the operators are linear or if strong assumptions on the distribution of the input data are made. A versatile approach to estimating the influence of errors in the input and to investigate the sensitivity to parameters arises from the application of UQ to the PDEs that are used in computer vision, image processing, and image analysis. Together with a suitable Ansatz space for the random fields that are used in the SPDEs, this approach results in corresponding stochastic counterparts of the classical PDEs.

To give more detail, let us assume pixels or voxels in images are identified with random variables yielding random fields. Such images are then called *stochastic images*. When represented with a spectral approach such the Wiener-Askey polynomial chaos, the solution of the SPDE does not describe a classical image but rather the modes of the stochastic expansion. Classical images for the moments of the image's distribution are obtained from the modes by further computations.

For sensitivity analysis, our concept lies in replacing parameters by random variables. Again, this leads to stochastic PDEs and random fields as their solutions that show the influence of parameters on the results of computer vision operators.

CHAPTER 2

Partial Differential Equations and Their Numerics

Partial differential equations (PDEs) are among the most useful tools in mathematical modeling. Applications range from the description of physical processes over chemical, biological, and biomedical applications to the modeling of financial markets and production processes. In the natural sciences, PDEs often originate from fundamental principles such as conservation laws. Also, the principle of energy minimization leads to PDEs as the determining equations.

In the past few decades, modeling with PDEs has also influenced research in computer graphics, image processing, and computer vision. Thereby, images are identified with physical quantities. A PDE-based computer vision operator takes such an image as parameter or initial data, and it yields a processed version of the original image as the solution of a PDE.

A review of prominent PDE-based image processing operators is given in Chapter 3. Here, we review some basics of PDEs and their numerical treatment. Our discussion is based on the classical diffusion equation, also known as the heat equation, and its steady state, Poisson equation. These equations have been studied in the context of computer vision tasks (see Sections 3.2 and 3.3), and for us, they serve as illustrative examples explaining the basic principles of the numerical treatment of PDEs.

The underlying physical principle for the diffusion equation is the so-called *continuity equation*, which states that a conserved quantity (such as mass, energy, momentum, etc.) u that is observed in a closed system $D \subset \mathbb{R}^d$ changes according to the divergence of a flux j of this quantity, i.e.,

$$\partial_t u(t, x) + \mathrm{div}(j(t, x)) = 0.$$

In this PDE, the quantity u and its flux depend on time t and spatial location $x \in D$, i.e., $u : \mathbb{R}^+ \times D \to \mathbb{R}$ and $j : \mathbb{R}^+ \times D \to \mathbb{R}^d$.

A more general form of the continuity equation takes into account that the quantity u may be produced or consumed inside D. In the equation, the net rate of production and consumption, i.e., the difference between *sources* and *sinks*, is added to the right-hand side, thus

$$\partial_t u(t, x) + \mathrm{div}(j(t, x)) = f(t, x). \tag{2.1}$$

The second ingredient for the derivation of the diffusion equation is the so-called *Fick's law*, which states that in diffusion of a quantity, the flux is proportional to the quantity's gradient, but pointing in the opposite direction, i.e.,

$$j = -w\nabla u. \tag{2.2}$$

The constant of proportionality w is called the *diffusion coefficient*. In the case of an isotropic diffusion, i.e., in a material that does not prefer diffusion in a particular direction, it is a positive real number $w \in \mathbb{R}^+$. In the case of anisotropic diffusion, i.e., in a material that has preferred diffusion directions (e.g., because of a layered structure), it is a matrix $w \in \mathbb{R}^{d \times d}$. It is possible that the diffusion coefficient depends on the spatial location $x \in D$. Also, it may exhibit some dynamics, thus $w = w(t, x)$.

Combining Fick's law (2.2) and the continuity equation (2.1) yields the famous *diffusion equation*, also known as the *heat equation*:

$$\partial_t u(t, x) - \text{div}(w(t, x)\nabla u(t, x)) = f(t, x). \tag{2.3}$$

We will see this equation again in Chapter 3.2 as it is one of the prominent examples for a computer vision operator.

A fundamental difference between differential equations and algebraic equations (such as a simple quadratic equation) is that a differential equation such as (2.3) may have infinitely many solutions. In fact, to select one unique solution from all possible functions that satisfy (2.3), we have to specify how the solution shall behave at the boundary of the computational domain, which is the space-time cylinder $\mathbb{R}^+ \times D$. The boundary has two components, the "bottom" of the cylinder $\{0\} \times D$ and the "side" of the cylinder $\mathbb{R}^+ \times \partial D$.

At $\{0\} \times D$, the value of the solution is prescribed as some *initial condition* $u_0 \colon D \to \mathbb{R}$, i.e.,

$$u(0, x) = u_0(x) \quad \text{for } x \in D. \tag{IC}$$

Physically, this serves as the state of the quantity u as it was when the observation started. At $\mathbb{R}^+ \times \partial D$, various conditions may be prescribed. The two most widely used *boundary conditions* (BC) are

$$
\begin{array}{llll}
u(t, x) = 0 & \text{for } (t, x) \in \mathbb{R}^+ \times \partial D & \text{Dirichlet or essential BC,} & \text{(DBC)} \\
\nabla u(t, x) \cdot \nu(x) = 0 & \text{for } (t, x) \in \mathbb{R}^+ \times \partial D & \text{Neumann or natural BC,} & \text{(NBC)}
\end{array}
$$

where $\nu(x)$ is the outward pointing normal vector to the boundary ∂D. The Neumann BC (NBC) is also referred to as a *no-flux* BC, as it models the fact that the quantity u is not allowed to diffuse across the boundary of D, just as if the system were insulated. Note that for both versions of BC, inhomogeneous variants with nonvanishing right-hand sides are possible as well. Also, the boundary ∂D may be split into several parts in which one or the other BC is prescribed (mixed BC).

To derive our template PDE, we combine (2.3), (IC), and (NBC) to arrive at the *initial-boundary value problem*:

For given $f, w : \mathbb{R}^+ \times D \to R$, and $u_0 : D \to \mathbb{R}$, find $u : \mathbb{R}^+ \times D \to \mathbb{R}$ such that

$$
\begin{aligned}
\partial_t u(t, x) - \operatorname{div}(w(t, x)\nabla u(t, x)) &= f(t, x) && \text{for } (t, x) \in \mathbb{R}^+ \times D, \\
u(0, x) &= u_0(x) && \text{for } x \in D, \\
\nabla u(t, x) \cdot v(x) &= 0 && \text{for } (t, x) \in \mathbb{R}^+ \times \partial D.
\end{aligned}
\tag{PE}
$$

This is our prototype of a *parabolic* second-order PDE.

In some applications, it is of interest to study the steady state of the evolution of u under the heat equation. If such a steady state exists, the temporal change $\partial_t u$ would vanish in (2.3). Assuming that the diffusion coefficient and the sources/sinks do not depend on time, the removal of the change in time results in the so-called *Poisson equation*:

$$
-\operatorname{div}(w(x)\nabla u(x)) = f(x).
$$

Combining this equation with a Dirichlet BC (cf. (DBC)), we arrive at the *boundary value problem*.

For given $f, w : D \to \mathbb{R}$, find $u : D \to \mathbb{R}$ such that

$$
\begin{aligned}
-\operatorname{div}(w(x)\nabla u(x)) &= f(x) && \text{for } x \in D, \\
u(x) &= 0 && \text{for } x \in \partial D.
\end{aligned}
\tag{EE}
$$

This is our prototype of an *elliptic* second-order PDE.

For the Dirichlet BC (DBC) is easy to consider an inhomogeneous version

$$
u(t, x) = g(x) \quad \text{for } (t, x) \in \mathbb{R}^+ \times \partial D
\tag{IDBC}
$$

for some given function $g : D \to \mathbb{R}$. In fact, defining $\tilde{u} := u - g$ and applying the differential operator $-\operatorname{div}(w\nabla)$ we find that $\tilde{u}$ must solve

$$
\begin{aligned}
-\operatorname{div}(w(x)\nabla \tilde{u}(x)) &= f(x) + \operatorname{div}(w(x)\nabla g(x)) && \text{for } x \in D, \\
\tilde{u}(x) &= 0 && \text{for } x \in \partial D.
\end{aligned}
$$

So, $\tilde{u}$ is the solution to the Possion equation with a modified right-hand side and homogeneous Dirichlet BC.

Note that we do not discuss conditions on f, w, u_0 that guarantee the existence of solutions. However, we need to acknowledge that for most application problems, equations (PE) and (EE) do not exhibit solutions that can be expressed analytically. Instead, it is necessary to solve the equations numerically, thereby approximating the true unknown solutions. The central goal of numerical approaches is to *discretize* the equations in order to transform the problem over infinite-dimensional spaces into a problem where solutions u lie in finite-dimensional spaces that can be treated by means of linear algebra. The discretization process starts with a discretization of the computational domain, continues with a discretization of the differential operators, and arrives at a system of algebraic equations. In the sections to follow, we describe two of the most widely used methods for the computational treatment of our template equations. We will not discuss these approaches in all their generality but instead will focus on aspects that are relevant for our applications in computer vision.

2.1   SOME BASICS FROM FUNCTIONAL ANALYSIS

To analyze (computer vision) models or the behavior of numerical methods used to discretize such models, it is convenient to utilize the rich toolbox of functional analysis. We now briefly review some of the main notions and function spaces to be used in the subsequent sections.

Assuming that the reader is familiar with the notion of vector spaces, we start with the definition of the three most important tools for analysis. Note that, in the context of computer vision, we work with real-valued vector spaces. Of course it is also possible to work with complex-valued vector spaces, which might make sense in signal processing applications outside the scope of this work.

Definition 2.1   Let $\mathcal{V}$ denote a real-valued vector space. A *metric* (also called *distance*) is a map $d(\cdot, \cdot) : \mathcal{V} \times \mathcal{V} \to \mathbb{R}_0^+$ that fulfills the following properties:

1. For all $u, v \in \mathcal{V}$: $d(u, v) \geq 0, = 0$ if and only if $u = v$ (*positive definiteness*).

2. For all $u, v \in \mathcal{V}$: $d(u, v) = d(v, u)$ (*symmetry*).

3. For all $u, v, w \in \mathcal{V}$: $d(u, w) \leq d(u, v) + d(v, w)$ (*sub-additivity or triangle inequality*).

A vector space $\mathcal{V}$ that is equipped with a metric d, i.e., the pair $(\mathcal{V}, d)$, is called *metric space*.

Definition 2.2   Let $\mathcal{V}$ denote a real-valued vector space. A *norm* is a function $\| \cdot \| : \mathcal{V} \to \mathbb{R}$ that fulfills the following properties:

1. For all $v \in \mathcal{V}$: $\|v\| \geq 0, = 0$ if and only if $v = 0$ (*positive definiteness*).

2. For all $a \in \mathbb{R}, v \in \mathcal{V}$: $\|av\| = |a| \, \|v\|$ (*absolute homogeneity*).

3. For all $u, v \in \mathcal{V}$: $\|u + v\| \leq \|u\| + \|v\|$ (*sub-additivity or triangle inequality*).

Thus, a norm assigns a real number to each element (vector, function) of $\mathcal{V}$. A vector space that is equipped with a norm $(\mathcal{V}, \|\cdot\|)$ is called a *normed space*. Since a norm induces a metric by the definition $d(u, v) := \|u - v\|$, every normed space is also a metric space. Well-known norms are the Euclidean norms or Hölder-norms on the vector spaces $\mathbb{R}^d$. In the context of PDEs, however, we will mostly work with vector spaces that contain functions. Thus, in this context, a norm assigns a real number to a function.

Definition 2.3   Let $\mathcal{V}$ denote a real-valued vector space. An *inner product* (also called *scalar product*) is a map $(\cdot, \cdot) : \mathcal{V} \times \mathcal{V} \to \mathbb{R}$ that fulfills the following properties:

1. For all $v \in \mathcal{V}$: $(v, v) \geq 0, = 0$ if and only if $v = 0$ (*positive definiteness*).

2. For all $u, v \in \mathcal{V}$: $(u, v) = (v, u)$ (*symmetry*).

3. For all $a \in \mathbb{R}, u, v, w \in \mathcal{V}$: $(au + v, w) = a(u, w) + (v, w)$ (*linearity*).

It is important to note that an inner product $(\cdot, \cdot)$ naturally defines a norm $\|\cdot\|$ through the definition $\|v\| = \sqrt{(v, v)}$. Consequently, an *inner product space* $(\mathcal{V}, (\cdot, \cdot))$ is also a normed space and thus a metric space. In inner product spaces, we have the useful *Cauchy Schwarz inequality*, i.e., for all $u, v \in \mathcal{V}$

$$|(u, v)| \leq \|u\| \cdot \|v\|.$$

The three important tools of distance, norm, and inner product help us in analyzing PDE-based computer vision models and their discretizations. One of the fundamental concepts related to this is convergence.

Definition 2.4   Let $\mathcal{V}$ be a metric space with metric d. A sequence (v_n) of elements of $\mathcal{V}$ *converges* to $v \in \mathcal{V}$ if the sequence of real numbers $d(v_n, v)$ converges to zero as n tends to infinity. In such a case, we call v the *limit* of v_n as n tends to infinity and we write $v_n \to v$ as $n \to \infty$.

In other words, a sequence of vectors/functions from the metric space $\mathcal{V}$ converges to a limit if the distance of the members of the sequence to that limit goes to zero as the index n goes to infinity.

Definition 2.5   Let $\mathcal{V}$ be a metric space with metric d. A sequence (v_n) of elements of $\mathcal{V}$ is called a *Cauchy sequence* if for all $\varepsilon > 0$ there is a threshold index N such that for $n, m > N$, the members of the sequence come closer than ε, i.e., $d(v_n, v_m) < \varepsilon$.

Thus, roughly speaking, in a Cauchy sequence, the members of the sequence come closer and closer together with a larger and larger index. This suggests that the sequence should converge to a limit, which, however, may not exist in the space $\mathcal{V}$. The spaces $\mathcal{V}$ for which the limit exists inside $\mathcal{V}$ are particularly important.

Definition 2.6   A metric space $\mathcal{V}$ in which every Cauchy sequence converges to a limit in $\mathcal{V}$ is called *complete*.

In this context, the notion of completeness is combined with the above mentioned notions of spaces, i.e., a complete metric space is called a *Cauchy space*, a complete normed space is called a *Banach space*, and a complete inner product space is a *Hilbert space*.

Banach spaces and Hilbert spaces, and their structure in terms of inner products, norms, and metrics, are the most powerful mathematical objects in the analysis of PDE-based computer vision models. Therefore, in the remainder of this section we will review some instances of these spaces that will play an important role in the remainder of this book.

- The space of continuous functions (Banach space):

$$C^0(D) := \{v : D \to \mathbb{R} \,|\, v \text{ is continuous} \}, \quad \|v\|_{C^0(D)} := \|v\|_{\infty,D} = \sup_{x \in D} |v(x)|.$$

- The spaces of continuously differentiable functions for $k > 0$ (Banach spaces):

$$C^k(D) := \{v : D \to \mathbb{R} \,|\, v \text{ is } k\text{-times continuously differentiable}\},$$

$$\|v\|_{C^k(D)} := \sum_{|\alpha|=0}^{k} \|\partial_\alpha v\|_{\infty,D},$$

where we use a multi-index notation ∂_α to denote all possible partial derivatives of degree $|\alpha|$. We denote the space of functions that are infinitely often continuously differentiable by $C^\infty(D)$. Such functions are called *smooth*.

- The *Lebesque spaces* for $p \in [1, \infty)$ (Banach spaces, Hilbert space for $p = 2$):

$$L^p(D) := \{v : D \to \mathbb{R} \,|\, \|v\|_{p,D} < \infty\}, \quad \|v\|_{p,D} := \left(\int_D |v(x)|^p \, dx \right)^{\frac{1}{p}} .$$

For $p = \infty$ we can define $L^\infty(D) := \{v : D \to \mathbb{R} \mid \|v\|_{\infty,D} < \infty\}$, where $\|v\|_{\infty,D} :=$ ess $\sup_{x \in D} |v(x)|$ is the essential supremum. The space $L^2(D)$ is a Hilbert space with an inner product $(u, v)_{2,D} := \int_D u(x)v(x)\,dx$.

- The *Sobolev spaces* for $p \in [1, \infty)$ and $k > 0$ (Banach spaces, Hilbert spaces for $p = 2$):

$$W^{k,p}(D) := \{v : D \to \mathbb{R} \mid \|v\|_{k,p,D} < \infty\}, \quad \|v\|_{k,p,D} := \left(\sum_{|\alpha|=0}^{k} \|\partial_\alpha v\|_{p,D}^p\right)^{\frac{1}{p}},$$

where we use again a multi-index notation ∂_α to denote partial derivatives of degree $|\alpha|$. In the context of Sobolev spaces, however, derivatives are meant in the *weak* or *distributional* sense (in contrast to the C^k spaces). We say that a function $v : D \to \mathbb{R}$ is a weak α^{th} partial derivative of $u : D \to \mathbb{R}$ if for all smooth functions $\phi \in C^\infty(D)$ with compact support, we have

$$\int_D u\partial_\alpha \phi \, dx = (-1)^{|\alpha|} \int_D v\phi \, dx.$$

Note that in case the function u has a continuous classical derivative, it coincides with the weak derivative. However, weak derivatives may also exist in cases where classical derivatives are not defined. For the case of $p = 2$, we define the inner product as $(u, v)_{k,2,D} := \sum_{|\alpha|=0}^{k} (\partial_\alpha u, \partial_\alpha v)_{2,D}$. Also for $p = 2$, we use the notation $H^k := W^{k,2}$.

Note that for the Lebesque space L^2, it is possible to use measures in the integration other than the Lebesque measure dx. In particular, one can use probability measures and thus define a Hilbert space structure on a probability space. We will use this in Chapter 4 when we discuss the numerics of stochastic PDEs.

2.2   FINITE DIFFERENCE METHOD

In view of computer vision applications, let us assume that the domain is rectangular, i.e., $D = [a, b] \times [c, d] \subset \mathbb{R}^2$ such that we can overlay a regular and structured grid of smaller rectangles that discretize the domain. To this end, for some $N_1, N_2 \in \mathbb{N}$, we introduce step-sizes $h_1 = \frac{b-a}{N_1}$, and $h_2 = \frac{d-c}{N_2}$, which represent the distance of gridlines in the horizontal and vertical directions. The horizontal and vertical gridlines intersect at nodes

$$(a + i_1 h_1, c + i_2 h_2) \in D, \quad \text{for } i_1 = 0, \ldots, N_1, \text{ and } i_2 = 0, \ldots, N_2 \tag{2.4}$$

as shown in Figure 2.1.

With this discretization of the domain D into the set of $N := (N_1 + 1)(N_2 + 1)$ nodes (2.4), we can transfer the PDE (EE) into N equations for the values of the solution u at the nodes. Here we have to take into account that the BC of (EE) already prescribes values

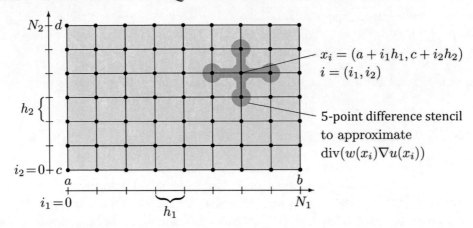

Figure 2.1: The discretization of the domain D with a uniform grid of rectangular elements.

at the boundary nodes, thus we arrive at

$$-\mathrm{div}(w(x)\nabla u(x)) = f(x) \quad \text{for } x = (a + i_1 h_1, c + i_2 h_2), \text{for } i_1 = 1, \ldots, N_1 - 1,$$
$$\text{and } i_2 = 1, \ldots, N_2 - 1,$$
$$u(x) = 0 \quad \text{for } x = (a + i_1 h_1, c + i_2 h_2), \text{for } i_1 = 0, N_1,$$
$$\text{and } i_2 = 0, N_2. \tag{2.5}$$

Next we have to take care of the differential operators div and ∇ and explain how to evaluate these on the basis of the values of u at the grid nodes. To ease notation, in the following, we will denote the evaluation of a function at node $(a + i_1 h_1, c + i_2 h_2)$ with a superscript i_1, i_2, e.g.,

$$u^{i_1,i_2} = u(a + i_1 h_1, c + i_2 h_2), w^{i_1,i_2} = w(a + i_1 h_1, c + i_2 h_2), f^{i_1,i_2} = f(a + i_1 h_1, c + i_2 h_2).$$

In the *Finite Difference Method* (FDM), partial derivatives are approximated by difference quotients that use values at the grid nodes. The basic difference quotients are

$$\partial_1 v(a + i_1 h_1, c + i_2 h_2) \approx D_{1,+}^{i_1,i_2} v := \frac{v^{i_1+1,i_2} - v^{i_1,i_2}}{h_1} \qquad \text{forward difference,}$$

$$\partial_1 v(a + i_1 h_1, c + i_2 h_2) \approx D_{1,-}^{i_1,i_2} v := \frac{v^{i_1,i_2} - v^{i_1-1,i_2}}{h_1} \qquad \text{backward difference,} \tag{2.6}$$

$$\partial_1 v(a + i_1 h_1, c + i_2 h_2) \approx D_{1,\circ}^{i_1,i_2} v := \frac{v^{i_1+1,i_2} - v^{i_1-1,i_2}}{2h_1} \qquad \text{central difference}$$

for some function $v : D \to \mathbb{R}$ and with analogous expressions for difference quotients in the second coordinate direction, i.e.,

$$\partial_2 v(a + i_1 h_1, c + i_2 h_2) \approx D_{2,+}^{i_1,i_2} v := \frac{v^{i_1,i_2+1} - v^{i_1,i_2}}{h_2}.$$

. . .

The forward/backward difference quotients are of first order, i.e., they have an error of $\mathcal{O}(h_i)$, and the central difference quotient has an error of $\mathcal{O}(h_i^2)$, as can be seen easily from a Taylor series expansion of v. Note that it is straightforward to use other difference schemes that yield higher-order approximations.

With the difference quotients, we can approximate the differential operators in (EE). To this end, we use a forward and a backward difference to approximate $w(x)\nabla u(x)$ and a difference between these two to approximate the further derivative from the div-operator. Thus, for $x = (a + i_1 h_1, c + i_2 h_2)$,

$$
\begin{aligned}
\operatorname{div}(w(x)\nabla u(x)) &= \partial_1(w(x)\partial_1 u(x)) + \partial_2(w(x)\partial_2 u(x)) \\
&\approx \frac{1}{h_1}\left(w^{i_1+1,i_2} D_{1,+}^{i_1,i_2}u - w^{i_1-1,i_2} D_{1,-}^{i_1,i_2}u\right) + \\
&\quad \frac{1}{h_2}\left(w^{i_1,i_2+1} D_{2,+}^{i_1,i_2}u - w^{i_1,i_2-1} D_{2,-}^{i_1,i_2}u\right) \\
&= \frac{1}{h_1^2}\left(w^{i_1+1,i_2}u^{i_1+1,i_2} - (w^{i_1+1,i_2} + w^{i_1-1,i_2})u^{i_1,i_2} + w^{i_1-1,i_2}u^{i_1-1,i_2}\right) \\
&\quad \frac{1}{h_2^2}\left(w^{i_1,i_2+1}u^{i_1,i_2+1} - (w^{i_1,i_2+1} + w^{i_1,i_2-1})u^{i_1,i_2} + w^{i_1,i_2-1}u^{i_1,i_2-1}\right).
\end{aligned}
$$

This formula for approximating the value of the differential operators at node i_1, i_2 involves the values of w and u at the four neighboring grid nodes $i_1 \pm 1, i_2 \pm 1$ and the node itself. Using the :-scalar product on matrices, $A : B = \sum_{jk} a_{jk}b_{jk}$, i.e., entry-wise multiplication and summation, we can express the discretization conveniently as

$$
\operatorname{div}(w(x)\nabla u(x)) \approx S^{i_1,i_2} : U^{i_1,i_2} \quad \text{for } x = (a + i_1 h_1, c + i_2 h_2),
$$

where the 3×3 matrices S^{i_1,i_2} and U^{i_1,i_2} are defined as

$$
S^{i_1,i_2} := \begin{pmatrix}
0 & \frac{w^{i_1,i_2-1}}{h_2^2} & 0 \\
\frac{w^{i_1-1,i_2}}{h_1^2} & -\frac{w^{i_1-1,i_2}+w^{i_1+1,i_2}}{h_1^2} - \frac{w^{i_1,i_2-1}+w^{i_1,i_2+1}}{h_2^2} & \frac{w^{i_1+1,i_2}}{h_1^2} \\
0 & \frac{w^{i_1,i_2+1}}{h_2^2} & 0
\end{pmatrix}
$$

$$
U^{i_1,i_2} := \begin{pmatrix}
u^{i_1-1,i_2-1} & u^{i_1,i_2-1} & u^{i_1+1,i_2-1} \\
u^{i_1-1,i_2} & u^{i_1,i_2} & u^{i_1+1,i_2} \\
u^{i_1-1,i_2+1} & u^{i_1,i_2+1} & u^{i_1+1,i_2+1}
\end{pmatrix}.
$$

Using this notation, we can turn (2.5) into a fully discrete version:

$$
\begin{aligned}
-S^{i_1,i_2} : U^{i_1,i_2} &= f^{i_1,i_2} \quad \text{for } i_1 = 1, \ldots, N_1 - 1, i_2 = 1, \ldots, N_2 - 1 \\
u^{i_1,i_2} &= 0 \qquad \text{for } i_1 = 0, N_1, \text{ and } i_2 = 0, N_2.
\end{aligned} \tag{2.7}
$$

This is in fact a linear system of equations in the unknowns u^{i_1,i_2}. It can be solved by any of the standard direct or iterative solvers for linear systems of equations, e.g., Jacobi, Gauss-Seidel, or

Conjugate Gradient (CG) method. Note that in case of a constant diffusion coefficient $w \equiv 1$ and for $h_1 = h_2$, we get back the well-known 5-point stencil of the Laplace operator, independent of the location i_1, i_2 in the grid

$$
S = \frac{1}{h^2} \begin{pmatrix} & 1 & \\ 1 & -4 & 1 \\ & 1 & \end{pmatrix}.
$$

To understand the structure of the system of equations (2.7), we reorder the unknowns in the classical way of traversing a regular grid with quadrilateral elements: from left to right, from bottom to top (i.e., *lexicographically*). Thus, we introduce vectors

$$
U = (u^{0,0}, \ldots, u^{N_1,0}, u^{0,1}, \ldots, u^{N_1,N_2-1}, u^{0,N_2}, \ldots, u^{N_1,N_2}) \in \mathbb{R}^N
$$

and analog for the right-hand-side F. In this ordering, we have one index i running from 1 to $N = (N_1 + 1)(N_2 + 1)$. Of course, there is a one-to-one correspondence between this index i and the coordinates i_1, i_2 in the grid.

Our goal is now to assemble all equations from (2.7) into one large system of equations, such that we can write it as $LU = F$. Collecting all local equations from the difference stencils of (2.7) into a global matrix (and, for the time being, not respecting the Dirichlet BC) yields the 5-band matrix

$$
L = \begin{pmatrix}
\tilde{A}_1 & * & B_1^+ & & & & \\
* & A_2 & * & B_2^+ & & & \\
\ddots & \ddots & \ddots & \ddots & \ddots & & \\
& B_i^- & * & A_i & * & B_i^+ & \\
& & \ddots & \ddots & \ddots & \ddots & \ddots \\
& & & B_{N-1}^- & * & A_{N-1} & * \\
& & & & B_N^- & * & \tilde{A}_N
\end{pmatrix} \in \mathbb{R}^{N \times N} \tag{2.8}
$$

in which

$$
A_i = \left[\frac{w^{i_1-1,i_2}}{h_1^2}, \quad -\frac{w^{i_1-1,i_2} + w^{i_1+1,i_2}}{h_1^2} - \frac{w^{i_1,i_2-1} + w^{i_1,i_2+1}}{h_2^2}, \quad \frac{w^{i_1+1,i_2}}{h_1^2} \right],
$$

$$
B_i^- = \frac{w^{i_1,i_2-1}}{h_2^2}, \qquad B_i^+ = \frac{w^{i_1,i_2+1}}{h_2^2},
$$

and where $*$ represents $N_1 - 1$ bands of zeros.

For all nodes i that are located at the boundary, we need modified versions $\tilde{A}_i$, $\tilde{B}_i^-$ and $\tilde{B}_i^+$ in the matrix: on the one hand, we shall not run "out of bounds" with the coordinates i_1 and i_2. On the other hand, for the later use in the FDM discretization of (PE), it is convenient to assume that there would be a layer of *ghost nodes* outside the domain, i.e., at $(a + i_1 h_1, c + i_2 h_2)$ for $i_1 = -1, N_1 + 1$ and $i_2 = -1, N_2 + 1$. The values at these nodes are fixed to the same value as the closest boundary node. Consequently, any difference between them vanishes and need not be considered in the difference stencil of the boundary node. For example, for $i_1 = i_2 = 0$, i.e., for the first node in the lower left corner of the grid, we have

$$\tilde{A}_1 = \left[-\frac{w^{i_1+1,i_2}}{h_1^2} - \frac{w^{i_1,i_2+1}}{h_2^2}, \quad \frac{w^{i_1+1,i_2}}{h_1^2} \right].$$

Analog adjusted versions of A_i and $B_i^{\pm}$ are needed for nodes at the other corners and vertical or horizontal boundary. It is left to the reader to explore the respective formulas.

The matrix L from (2.8) is a 5-band matrix, thus, most of its entries are zero. In its current form, it represents natural (Neumann) BC (NBC) as we will need them for the discretization of problem (PE). To represent the actual Dirichlet BC of (2.7), we need a further modification of L. To this end, we have to introduce the equations $u^i = 0$ into the system $LU = F$. This is achieved by erasing all lines and columns of L that correspond to the index of a boundary node. Next, a value of 1 is placed on the diagonal of these lines. Finally, the corresponding entry of vector F is replaced with a 0. The structure of the adjusted matrix L is shown in Figure 2.2.

Finally, we are able to express (PE) as a fully discrete problem in the form

$$LU = F.$$

The matrix L is symmetric and positive definite, which makes the linear system of equations a candidate for an iterative method like conjugate gradients. Note that the above explanations can be repeated straightforwardly for the case of 3D images or even higher dimensions. The difference is that the resulting matrix L will have a slightly more complicated structure.

2.3   WEAK FORMULATION

The FDM discussed in the previous paragraph is easy to understand and simple to implement. For computer vision applications, the FDM approach works particularly well since domains D are rectangular or cuboid.

If one is interested in computer vision applications on surfaces (manifolds), it becomes necessary to go over to unstructured grids on which the use of finite difference approximations becomes difficult or even impossible.

Also, in view of the analysis of the models, one must note that the FDM is based on the classical formulation (EE) and (PE), which assumes that all derivatives involved in the equation exist in the classical sense, here $u \in C^2(D)$ and $w \in C^1(D)$.

It is known, however, that when the diffusion coefficient a is only piecewise continuous, the solutions may have kinks—places at which classical derivatives do not exist. In fact, such situations of piecewise continuous diffusion coefficients appear often in PDE-based computer

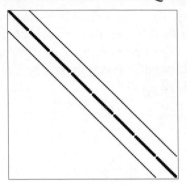 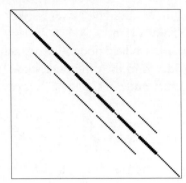

Figure 2.2: The sparsity structure of the matrix resulting from the FDM on a regular 2D grid is shown, i.e., every nonzero entry in the matrix is represented by a black dot. The original matrix L (cf. (2.8)) is shown on the left, whereas on the right, its adjustment to Dirichlet BC is depicted. A block structure of the matrices is clearly visible, where every block corresponds to one line of the grid. Here, the grid has dimension $N_1 = 16$ and $N_2 = 8$, i.e., we have 8×8 blocks each of size 16×16. In the adjusted matrix (on the right) the identity-rows that represent the Dirichlet BC can be easily identified.

vision. As a consequence of these observations, we present an alternative formulation of the PDEs, the so-called *weak formulation*, and utilize the finite element method (FEM) instead of FDM in Section 2.4.

To derive the weak formulation of (EE), we multiply the equation by a *test function* v and integrate over the domain D to obtain

$$- \int_D \operatorname{div}(w(x)\nabla u(x))\, v(x)\, dx = \int_D f(x)\, v(x)\, dx.$$

Using integration by parts (also known as Green's first identity), we conclude from this that

$$\int_D w(x)\nabla u(x) \cdot \nabla v(x)\, dx - \int_{\partial D} w(x)v(x)\nabla u(x) \cdot v(x)\, ds = \int_D f(x)\, v(x)\, dx$$

for any test function v. If u fulfills a Neumann BC (NBC), the boundary integral will vanish. If u is supposed to fulfill a Dirichlet BC (DBC), we shall assume that the test function v also vanishes at the boundary with the consequence that the boundary integral vanishes as well. Even if we prescribe a mixed BC to u, we can select the test function such that the boundary integral vanishes. This leaves us with

$$\int_D w(x)\nabla u(x) \cdot \nabla v(x)\, dx = \int_D f(x)\, v(x)\, dx \quad \text{for all test functions } v.$$

In this equation, derivatives do not need to exist in the classical sense. It is sufficient if they exist in a weak sense as defined above in the context of the Sobolev spaces. Thus, we assume that

$u \in H^1(D)$, i.e., u has weak derivatives of order 1. If u is supposed to fulfill (DBC), we need to restrict this space to

$$H_0^1(D) = \{u \in H^1(D) \,|\, u|_{\partial D} \text{vanishes}\}.$$

Note that the "vanishing" of u at the boundary of D must be interpreted in the sense of a *trace* since u may not be continuous at ∂D (see e.g., [31]).

Finally, we have arrived at the *weak formulation* of the problem (EE) for which we use the notation of the L^2 scalar product for the integrals, that is,

For given $f \in L^2(D)$ and $w \in L^\infty(D)$, find $u \in H_0^1(D)$ such that

$$(w\nabla u, \nabla v)_{2,D} = (f, v)_{2,D} \qquad \text{for all } v \in H_0^1(D). \qquad \text{(WEE)}$$

A solution of (WEE) is called the *weak solution* of (EE). The weak formulation has the advantage that only weak derivatives of u and v are required. If, however, the solution u exhibits classical derivatives, both formulations are equivalent, i.e., a weak solution is a classical solution and vice versa. Note that in the weak formulation, any BC is either built naturally into the equation (in the case of (NBC)) or constructed into the space (in the case of (DBC)). An analog weak formulation can be derived for (PE). In our expositions, however, we will introduce time-stepping first in Section 2.5 before transforming over to the weak form.

Sometimes it is convenient to denote the left-hand side of (WEE) with the bilinear form

$$A : H^1(D) \times H^1(D) \to \mathbb{R}, \quad A(u, v) = (w\nabla u, \nabla v)_{2,D}$$

and the right-hand side with the linear form

$$B : L^2(D) \to \mathbb{R}, \quad B(v) = (f, v)_{2,D},$$

such that the weak form (WEE) reads $A(u, v) = B(v)$ for all test functions $v \in H_0^1(D)$.

2.4 GALERKIN APPROACH AND FINITE ELEMENT METHOD

Without further approximation, the weak form (WEE) as stated is not accessible to numerical methods, because the Sobolev space $H_0^1(D)$ in which we search for the solution is infinite dimensional, i.e., there does not exist a basis with finitely many elements. To remedy this, the *Galerkin approach* transforms over to a finite-dimensional subspace $V_h \subset H_0^1(D)$ and looks for the best possible approximation to the solution in this space. If the space V_h is spanned by basis functions $P_i(x)$, i.e.,

$$V_h = \text{span}\{P_i \,|\, i \in \mathcal{I}\}$$

for some index set $\mathcal{I}$ with $N := |\mathcal{I}| < \infty$, we can express any function $v \in V_h$ as

$$v(x) = \sum_{i \in \mathcal{I}} v^i P_i(x), \quad \text{for } v^i \in \mathbb{R}. \qquad (2.9)$$

Sometimes the basis functions P_i are also referred to as *shape functions*.

> *Note that here and throughout this book, we use a superscript to indicate the index of the coefficient v^i for basis function P_i. We advise the reader to not confuse the superscript i of v^i with taking successive multiples of the variable, i.e., v to the power of i.*

When we replace $H_0^1(D)$ in (WEE) with $\mathcal{V}_h$, we can use the basis representation (2.9) on u. Furthermore, it suffices to *test* the weak form with all basis functions P_j. In fact, if (WEE) holds for all P_j, it will hold for any $v \in \mathcal{V}_h$ due to the linearity of the inner product $(\cdot, \cdot)_{2,D}$. So,

$$\left(w\nabla \left(\sum_{i \in \mathcal{I}} u^i P_i \right), \nabla P_j \right)_{2,D} = (f, P_j)_{2,D}, \quad \text{for all } j \in \mathcal{I}.$$

Using linearity of the inner product and the nabla operator, we obtain from this

$$\sum_{i \in \mathcal{I}} u^i (w\nabla P_i, \nabla P_j)_{2,D} = (f, P_j)_{2,D}, \quad \text{for all } j \in \mathcal{I}.$$

This is a discrete version of (WEE) in terms of a linear system of equations:

For given $f \in L^2(D)$, $w \in L^\infty(D)$ and basis functions $P_i, i \in \mathcal{I}$ find $U \in \mathbb{R}^N$ such that

$$LU = F, \qquad\qquad (\text{WEE}_h)$$

where $F \in \mathbb{R}^N$, $L \in \mathbb{R}^{N \times N}$ with $F_j = (f, P_j)_{2,D}$ and $L_{ij} = (w\nabla P_i, \nabla P_j)_{2,D}$.

Finite Element Method.   The final piece in the discretization of the weak formulation through the Galerkin approach is to determine appropriate basis functions P_i that span a space suitable for the problem under investigation. This procedure results in the *Finite Element Method* (FEM). To focus on computer vision applications, we assume, as for the FDM, that D is rectangular (or cuboid in three space dimensions) and that we introduce a regular grid of width h_1 and h_2 in the horizontal and vertical space direction. The intersection of these gridlines define the nodes of the discretized domain as shown in Figure 2.1. As a consequence of this discretization, we have partitioned the domain D into a set of smaller rectangular elements,

$$E_{i_1, i_2} = [a + i_1 h_1, a + (i_1 + 1)h_1] \times [c + i_2 h_2, c + (i_2 + 1)h_2],$$

for $i_1 = 0, \ldots, N_1 - 1, i_2 = 0, \ldots, N_2 - 1$ and analogously in three space dimensions. The union of all elements results in the whole domain D. The intersection between two elements is either the empty set, a common node or a common edge (or a common face in 3D). As before for the FDM, we order nodes in our grid such that we traverse them from left to right and bottom to top when being placed into the index set $\mathcal{I}$ from the Galerkin approach.

Central to the FEM is to define the set of interpolating basis functions P_i. It is convenient to use piecewise polynomials here to yield the approximation to the solution. In the simplest case, which is sufficient for the computer vision applications discussed in this book, we use piecewise bilinear (or trilinear in the 3D case) basis functions P_i, which fulfill a Lagrange interpolation property, i.e.,

$$\text{for all } i, j \in \mathcal{I} : P_i(x_j) = \delta_{ij}, \text{ and } P_i|_{E_j} \text{ is bilinear/trilinear,}$$

where δ_{ij} is the Kronecker δ. Thus, the space $\mathcal{V}_h$ will contain continuous and piecewise bilinear/trilinear functions. As a consequence of the Lagrange interpolation condition, the basis functions have small support that stretches over four or eight elements (bilinear/trilinear respectively) only. Figure 2.3 shows a basis function for the 2D case.

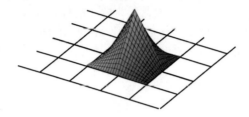

Figure 2.3: The graph of a piecewise bilinear basis function for the FEM discretization on a regular 2D grid with rectangles. Note that this function has kinks at the gridlines, thus it is not everywhere differentiable in the classical sense. Instead this basis function is weakly differentiable and contained in the Sobolev space $H^1(D)$.

To conclude the finite element discretization as one instance of the Galerkin approach, we need to discuss the entries of the vector F and the matrix L of (WEE$_h$). Note that in the context of the FEM the matrix L is called the *stiffness matrix*.

As a general computational trick, all integrals in the entries of L and F are computed by successively traversing all elements of the grid, i.e., $\int_D = \sum_{i \in \mathcal{I}} \int_{E_i}$. For each element E_i, the contributions to the various integrals in F and L can be accumulated into the corresponding locations of the vector and matrix. Moreover, only few elements will actually contribute to an integral, because the basis functions have small support.

Depending on the regularity of the functions f and w, we may utilize quadrature rules of sufficient exactness. For most of our applications, however, it turns out that *mass lumping* is possible: we use an inexact quadrature rule to compute F, for example by assuming that f is constant on each element, i.e.,

$$(f, P_j)_{2,D} = \int_D f(x) P_j(x) \, dx$$

$$= \sum_{E_i \cap \operatorname{supp} P_j \neq \emptyset} \int_{E_i} f(x) P_j(x) \, dx \approx \sum_{E_i \cap \operatorname{supp} P_j \neq \emptyset} f(c_{E_i}) \int_{E_i} P_j(x) \, dx,$$

where c_{E_i} is the center of the element E_i. Thus, we evaluate f at the center of the element E_i only, which can be considered as an averaging over E_i. Similarly, we may assume that the diffusion coefficient is constant per element, thus,

$$(w\nabla P_i, \nabla P_j)_{2,D} = \int_D w(x)\nabla P_i(x) \cdot \nabla P_j(x)\,dx$$
$$\approx \sum_{E_i \cap \mathrm{supp}(P_i P_j) \neq \emptyset} w(c_{E_i}) \int_{E_i} \nabla P_i(x) \cdot \nabla P_j(x)\,dx.$$

With this mass lumping, we earn some small additional smoothing of the solution. However, we gain an enormous simplification for the computations (and thus also for the implementation). Since all elements E_i of our grid are identical modulo translations, we can simply precompute all possible values of $\int_E P_j(x)\,dx$ and $\int_E \nabla P_i(x) \cdot \nabla P_j(x)\,dx$ on some reference element $E = [0, h_1] \times [0, h_2]$ (and analog for 3D) and store them in a lookup table (see Figure 2.4). Then, in a traversal of the grid to compute the entries of F and L, we recall the values from the lookup table, scale them with the averages of f and w inside of an element, and accumulate them into the corresponding locations of vector and matrix.

$$\hat{L} = \begin{pmatrix} \frac{2}{3} & -\frac{1}{6} & -\frac{1}{3} & -\frac{1}{6} \\ -\frac{1}{6} & \frac{2}{3} & -\frac{1}{6} & -\frac{1}{3} \\ -\frac{1}{3} & -\frac{1}{6} & \frac{2}{3} & -\frac{1}{6} \\ -\frac{1}{6} & -\frac{1}{3} & -\frac{1}{6} & \frac{2}{3} \end{pmatrix}, \qquad \hat{M} = \frac{1}{h_1 h_2} \begin{pmatrix} \frac{1}{9} & \frac{1}{18} & \frac{1}{18} & \frac{1}{36} \\ \frac{1}{18} & \frac{1}{9} & \frac{1}{36} & \frac{1}{18} \\ \frac{1}{18} & \frac{1}{36} & \frac{1}{9} & \frac{1}{18} \\ \frac{1}{36} & \frac{1}{18} & \frac{1}{18} & \frac{1}{9} \end{pmatrix}$$

Figure 2.4: Lookup tables $\hat{L}$ and $\hat{M}$ of values of the stiffness matrix and the mass matrix, respectively, for piecewise bilinear basis functions. Note that each row/column, and thus all entries of the reference stiffness matrix $\hat{L}$, sum up to zero, while all entries of the reference mass matrix sum up to one.

As a consequence of the local support of the basis functions P_i, most entries of the stiffness matrix L are zero, thus the matrix is sparse. In fact, for the regular grid that we have employed here, L will be a matrix with 9 bands (27 bands in 3D).

Dirichlet Boundary Conditions. It remains to discuss the incorporation of the BC of type given by (DBC). For the weak form, we had built these BC directly into the space $H_0^1(D)$. For the approximating subspace $V_h \subset H_0^1(D)$, it means that all basis functions P_i need to vanish at the boundary ∂D. Thus, for the FEM, basis functions located at boundary nodes, i.e., for $i_1 = 0, N_1, i_2 = 0, N_2$, do not carry degrees of freedom. Still they are contained in our current form of vector F and stiffness matrix L. Remember that the BC of type (NBC) occur naturally without any special construction.

Therefore, our final step in the FEM is an adjustment of L, analog to the modification for the FDM matrix, by eliminating all rows and columns that correspond to boundary nodes. A value 1 needs to be placed on the diagonal of these rows/columns. Also, the corresponding entries of vector F needs to be set to 0. With these modifications, we have introduced an algebraic equation $1 \cdot u_i = 0$ for each node i of the boundary, resulting in $u_i = 0$.

2.5   TIME-STEPPING SCHEMES

Up to now, we have only discussed the discretization of the elliptic boundary and initial value problem (EE). The discrete differential operators, i.e., the matrices that result from the FDM or FEM, will now be used in the discretization of the parabolic equation (PE). Roughly speaking, a parabolic equation is an elliptic equation plus time-dependence. With an appropriate time-discretization, we convert the parabolic equation into a sequence of elliptic equations, one for each time-step. These step-by-step elliptic problems are then solved with FDM or FEM.

To start, we introduce the time-step width $\tau > 0$ and define the time-points $t_l = l\tau$, for $l = 0, 1, 2, \ldots$. For a time-discrete version of (PE), we search to approximate the solution $u(t, x)$ at these time-points. As for the spatial discretization with FDM, key to the discretization is an approximation of the time-derivative $\partial_t u$ with the values of u at these time-points only. Among the multitude of possible time-stepping schemes, the following are the most popular in the context of PDEs:

$$\partial_t u(t_l, x) \approx \frac{1}{\tau}\big(u(t_{l+1}, x) - u(t_l, x)\big) \qquad \text{Explicit/Forward Euler Scheme} \qquad \text{(FE)}$$

$$\partial_t u(t_l, x) \approx \frac{1}{\tau}\big(u(t_l, x) - u(t_{l-1}, x)\big) \qquad \text{Implicit/Backward Euler Scheme} \qquad \text{(BE)}$$

$$\partial_t u(t_l, x) \approx \frac{1}{2\tau}\big(u(t_{l+1}, x) - u(t_{l-1}, x)\big) \qquad \text{Crank-Nicolson Scheme} \qquad \text{(CN)}$$

for any $x \in D$. Both (FE) and (BE) are first-order approximations to the time-derivative with an error of $\mathcal{O}(\tau)$. The (CN) scheme is a second-order approximation with an error $\mathcal{O}(\tau^2)$. These approximations are analog to the differences we have mentioned for the space-discretization in (2.6).

In the following, we will show the application of the explicit Euler scheme and its combination with the FDM. Also, we will discuss the application of the implicit Euler scheme and its combination with the FEM. All exploration of time-discretizations with the Crank-Nicolson scheme, and all other combinations between time-discretization and space-discretization, will be left to the reader.

Explicit Euler and FDM.   Applying the explicit Euler scheme to (PE) yields a semi-discrete (time-discrete) problem

$$u(t_{l+1}, x) = u(t_l, x) + \tau\big(\text{div}(w(t_l, x)\nabla u(t_l, x)) + f(t_l, x)\big), \quad \text{for } x \in D, l = 0, 1, \dots,$$
$$\nabla u(t_{l+1}, x) \cdot v(x) = 0 \qquad\qquad\qquad\qquad \text{for } x \in \partial D,$$
$$u(0, x) = u_0(x) \qquad\qquad\qquad\qquad \text{for } x \in D.$$

This shows why the scheme is called explicit: given the approximation $u(t_l, x)$ of u at some time-step t_l, we can compute the approximation at the following time-step t_{l+1} directly (explicitly) without inverting an equation.

To derive the fully discrete problem, we can use the objects from the FDM directly: let $U^l \in \mathbb{R}^N$ denote the vector of grid values of $u(t_l, x)$, $F^l \in \mathbb{R}^N$ denote the vector of grid values of $f(t_l, x)$, and let $L^l \in \mathbb{R}^{N \times N}$ denote the matrix (2.8) evaluated with the diffusion coefficient $w(t_l, x)$. Then, the fully discrete problem resulting from (FE) and FDM is

$$U^{l+1} = U^l + \tau L^l U^l + \tau F^l \qquad\qquad \text{for } l = 0, 1, \dots,$$

where $(U^0)_i = u_0(x_i)$ for all nodes of the grid. This is a very simple equation that involves a matrix vector product and some vector additions. The price one has to pay for this simplicity is a restriction on the values of τ for which the (FE) scheme is stable (i.e., numerical inaccuracies are not exaggerated over the course of the iteration). In general, the maximal possible time-step depends on the eigenvalues of the matrix L^l. In the case of the 2D diffusion equation (PE) and for $h_1 = h_2$, we have $\tau \leq h^2(4 \max w(t, x))^{-1}$. Thus, we have a coupling between spatial grid-width and temporal step size.

Note that this type of discretization is sometimes referred to as *forward time central space* (FTCS) scheme.

Implicit Euler and FEM.   Applying the implicit Euler scheme to (PE) yields a semi-discrete (time-discrete) problem of the form

$$u(t_l, x) - \tau \text{div}(w(t_l, x)\nabla u(t_l, x)) = u(t_{l-1}, x) + \tau f(t_l, x), \quad \text{for } x \in D, l = 1, 2, \dots,$$
$$u(0, x) = u_0(x) \qquad\qquad\qquad \text{for } x \in D.$$

To obtain $u(t_l, x)$ from given $u(t_{l-1}, x)$, an elliptic equation needs to be solved. Thus, $u(t_j, x)$ is defined implicitly through this equation. This is more elaborate; however, the implicit Euler scheme is unconditionally stable and works for all time-step sizes τ.

From here, we need to derive the weak form and utilize the Galerkin approach and the FEM to arrive at a fully discrete problem. As before, let $v \in H^1(D)$ be a test function. Note that for (PE), we work with natural boundary conditions (NBC) for which no extra condition in the space of test functions is needed. Multiplication of the equation with v and integration by parts yields

$$\int_D u(t_l, x)v(x)\, dx + \tau \int_D w(t_l, x)\nabla u(t_l, x) \cdot \nabla v(x)\, dx = \int_D (u(t_{l-1}, x) + \tau f(t_l, x))\, v(x)\, dx$$

for $l = 1, 2, \ldots$ as the weak form. With the expansion (2.9) and test functions $v(x) = P_j(x)$ as before, and by denoting with U^l the vector of unknown grid values of $u(t_l, x)$, we get the fully discrete form as

$$(M + \tau L^l)U^l = MU^{l-1} + \tau F^l, \quad \text{for } l = 1, 2, \ldots.$$

Here, again, $(U^0)_i = u_0(x_i)$ for all grid nodes x_i and

$$F^l \in \mathbb{R}^N, \qquad (F^l)_j = \int_D f(t_l, x) P_j(x)\, dx = (f(t_l, x), P_j(x))_{2,D},$$

$$M \in \mathbb{R}^{N \times N}, \quad M_{ij} = \int_D P_i(x)\, P_j(x),\, dx = (P_i(x), P_j(x))_{2,D},$$

$$L^l \in \mathbb{R}^{N \times N}, \quad (L^l)_{ij} = \int_D w(t_l, x)\, \nabla P_i(x) \cdot \nabla P_j(x)\, dx = (w(t_l, x)\nabla P_i(x), \nabla P_j(x))_{2,D}.$$

We had already encountered the stiffness matrix L^l in the FEM discretization of (EE). Here we have an additional matrix, the *mass matrix* M, which contains the integrals of products of pairs of basis functions over the computational domain. Again, this matrix can be computed conveniently by traversing over all elements of the grid and by taking advantage of the fact that all elements coincide modulo translations. Thus, we can use a lookup table on the reference element. For the case of a rectangular 2D domain and piecewise bilinear basis functions, the mass matrix on the reference element, i.e., the lookup table, is shown in Figure 2.4.

Note that the mass matrix M exhibits the same sparsity structure as the stiffness matrix. Also, to accelerate computations (but taking into account some additional smoothing), it is again possible to use an inexact quadrature rule (mass lumping, see above), which leads to a diagonal mass matrix. On the reference element, the lookup table for the lumped mass matrix is $\hat{M}_{\text{lumped}} = (h_1 h_2)^{-1}\text{diag}(1/4)$. Also note that since we have prescribed natural/Neumann BC (NBC) in (PE), we do not have to adjust the matrices M and L^l. Instead, the BC are met automatically (naturally).

In each step, the resulting system of equations can be solved conveniently with an iterative solver like the conjugate gradient method. Also, to accelerate the solution process, in particular for 3D applications, it is possible to employ grid-adaptivity. For the grid types used here, and in the context of image processing, this is discussed in [86].

Note 2.7 Readers interested in more information about PDEs and their numerics are referred to this literature:

- [31]: Excellent introduction of elliptic PDEs

- [87]: Numerical mathematics including FDM

- [48]: Details about FEM presented above

- [50]: Introduction to scientific computing and efficient, parallel implementations of the algorithms discussed above

CHAPTER 3

Review of PDE-Based Image Processing

Note 3.1 Readers not familiar with PDE-based image processing may consult the following books:

- [108]: Overview of how to use diffusion equations for image processing tasks

- [92]: Introduction to image analysis with PDEs

- [68]: Image registration with medical applications

This chapter gives a brief cross-sectional view of the broad subject of PDE-based computer vision and image processing with the purpose of familiarizing the reader with the models, equations, and filters that have analogs for uncertain gray values, which we discuss in later sections. We focus on only a small variety of operators that have been presented and discussed in the past. For educational purposes, we start with simple diffusion operators for image denoising and segmentation. We then address level set methods and phase field methods for image segmentation. We close the chapter with a brief overview of PDE-based optical flow methods for image sequences.

From the viewpoint of applications, PDE-based computer vision competes with other algorithms that might perform better in terms of speed. However, from the viewpoint of understanding models and algorithms, PDE-based models have a clear advantage since they are backed by the rich toolboxes of functional analysis and numerical analysis. With sophisticated discretizations and implementations, PDE-based approaches can be competitive in terms of speed. They can also be the basis for the development of other high-performance algorithms for which understanding the solid mathematical basis of the PDE context is valuable.

In this chapter, we refrain from a detailed discussion of discretization and numerical aspects of the models and filters presented. We assume that the reader is already acquainted with these topics or that the review chapter (Chapter 2) has laid a foundation for understanding the following material. For the reader interested in more details, we will include various references for further reading. Readers who are familiar with PDE-based computer vision and image processing may either briefly browse through the chapter to get acquainted with our notation or completely skip it and continue with the discussion of the numerics of stochastic PDEs in Chapter 4.

3.1   MATHEMATICAL REPRESENTATION OF IMAGES

Before a discussion of various computer vision methods, we introduce the mathematical framework in which we are operating. As we are dealing with PDE-based methods, the central notion of image is based on infinite-dimensional spaces. The finite-dimensional notion of digital image comes into play with the discretization of these spaces.

Definition 3.2   An *image* is a function $u : D \to \mathbb{R}$, $x \mapsto u(x)$ that maps from the image domain $D \subset \mathbb{R}^d$, $d \in \{2, 3\}$ to the real numbers $\mathbb{R}$.

Note that we consider that an image is defined on a continuous 2D or 3D space and thus it is defined by image values in infinitely many points. We are also assuming that we have infinitely many image values possible. We restrict the image values to be scalar, e.g., gray values; however, a generalization to color images, complex image values, or other image value spaces is in most cases straightforward and easy. Also, a generalization to higher-dimension image domains $d > 3$ is possible and might make sense for certain applications.

In the case where we restrict the image domain to a discrete finite set of points, we arrive at the notion of digital image.

Definition 3.3   A *digital image* u is a set of image intensities $\{u^i\}_{i \in \mathcal{I}} \in \mathbb{R}^{|\mathcal{I}|}$ that is defined on the set of pixels in 2D (or voxels in 3D) $x_i \in D \subset \mathbb{R}^d$, $d \in \{2, 3\}$ for all $i \in \mathcal{I}$.

Let us emphasize, at this point, that throughout the book we will use Latin superscripts to denote coefficients for discretization in physical space D, and Greek subscripts to denote coefficients for stochastic discretization. Functions that are continuous in physical space D carry Latin subscripts, and continuous stochastic (basis) functions (i.e., continuous random variables) carry Greek superscripts. We advise the reader to carefully track the meaning of superscripts so as not to confuse them with repeated multiplications, i.e., powers, of the respective quantities.

In most cases, the set of pixels/voxels will lie on the nodes of a regular Cartesian grid over some rectangular or quadrilateral domain (see Figure 3.1 and also Sections 2.2 and 2.4). We can thus also identify the set of image intensities as elements of $\mathbb{R}^{n_x \times n_y}$ or $\mathbb{R}^{n_x \times n_y \times n_z}$ where $n_x, n_y, n_z \in \mathbb{N}$ are the dimensions of the digital image.

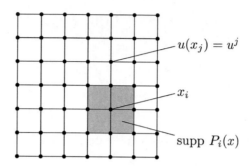

$u(x_j) = u^j$

x_i

supp $P_i(x)$

Figure 3.1: Sketch of the ingredients of a 2D digital image. A pixel is located at every intersection of the regular gridlines, and for every pixel, the corresponding FE basis function has its support in the elements around this pixel.

In addition to these definitions of image and digital image, it is important to define the notions of image sequence and its digital analog since image sequences play an important role in computer vision as well.

Definition 3.4   For some finite $T > 0$, an *image sequence* is a function $u : [0, T] \times D \to \mathbb{R}$, $(s, x) \mapsto u(s, x)$ that maps from the time-space cylinder $Q := [0, T] \times D, d \in \{2, 3\}$ to the real numbers $\mathbb{R}$.

Note that we use s to denote time here because t will be used later with a different meaning.
The definition of a digital image sequence is completely analogous to a digital image but with the difference that we use pairs of indices to enumerate the pixels/voxels in the sequence.

Definition 3.5   A *digital image sequence* is a set of intensities $\{u^i\}_{i \in \mathcal{I}} \in \mathbb{R}^{|\mathcal{I}|}$ defined on the set of time-space pixels/voxels $x_i \in Q$ for all time-space indices $i = (i_s, i_x) \in \mathcal{I}$. Of course, a digital image sequence can be seen as a set of digital images $\{u^{i_x}\}^{i_s}$.

Again, it is possible to identify a digital image sequence with an element from $\mathbb{R}^{n_s \times n_x \times n_y}$ or $\mathbb{R}^{n_s \times n_x \times n_y \times n_z}$, where n_s denotes the number of frames in the sequence. We will also need semi-digital images in which time or space is kept continuous:

Definition 3.6   A *semi-digital image sequence* is either

- a *space-digital image sequence*, i.e., a set of intensity functions $\{u^{i_x}(s)\}_{(i_s,i_x)\in\mathcal{I}}$, such that $u^{i_x} : [0, T] \to \mathbb{R}$, or

- a *time-digital image sequence*, i.e., a set of images $\{u^{i_s}(x)\}_{(i_s,i_x)\in\mathcal{I}}$, such that $u^{i_s} : D \to \mathbb{R}$.

The link between a continuous representation and a discrete representation of an image lies in

- *discretization*: transformation from an infinite-dimensional representation $u : D \to \mathbb{R}$ to a finite-dimensional digital image $\{u^i\}_{i\in\mathcal{I}}$, e.g., by the operation $u^i = u(x_i)$,

- *interpolation*: transformation from a digital image $\{u^i\}_{i\in\mathcal{I}}$ to a continuous representation in which the image values can be evaluated at infinitely many points of the domain D.

Of course, analog relationships hold for image sequences, which we do not repeat explicitly here and in the following.

The interpolation process can be connected with the notion of *shape functions* used in finite element discretizations of PDEs (see Section 2.4). Let P_i denote a piecewise bilinear (2D) or piecewise trilinear (3D) nodal basis function of the ith pixel/voxel, i.e., $P_i(x_j) = \delta_{ij}$ (Kronecker delta) for all $i, j \in \mathcal{I}$. Then a digital image $\{u^i\}$ is interpolated at every point x in the domain D by the image function

$$u(x) = \sum_{i\in\mathcal{I}} u^i P_i(x).\tag{3.1}$$

This representation of an image in terms of pixel/voxel-values and shape functions is connected with a bi-/trilinear interpolation rule. This rule is widely used, but of course, other shape functions and interpolation rules are also possible.

So far, we have not connected any regularity assumptions with the notion of image. With the toolbox of functional analysis available to us, we can utilize various tools as needed by the application, discretization, or theoretical analysis. A common assumption in the context of PDE-based image processing is that images lie in some Sobolev space, e.g., $u \in H^1(D)$, and have weak derivatives. Sometimes it is enough to assume that images are merely continuous, or square integrable, but sometimes further assumptions such as bounded variation and the like have to be made.

In PDE-based computer vision, (digital) images are considered as input data for PDEs, e.g., as the right-hand side, as coefficients, or as initial data for parabolic PDEs. The solution of the PDE is then considered to be the result of the computer vision operator. What PDE to choose depends on the specific computer vision task to be solved. In the remainder of this chapter, we will review some common computer vision tasks to prepare the reader for a discussion of their stochastic analogs in Chapter 6.

In many cases, the computer vision PDE is the result of the minimization of some energy functional. Writing the problem as a minimization either yields an elliptic PDE as the necessary condition for a minimum (Euler-Lagrange equation), or a parabolic PDE, if the search for an extremal point of the energy is conducted with a gradient descent. This approach directly connects to the notion of scale-space, been discussed extensively in the image processing community, see, e.g., [108]. Hyperbolic PDEs are used in computer vision filters with a dominant process that is front-tracking, such as in level set methods for image segmentation.

3.2   DIFFUSION FILTERING FOR DENOISING

In the computer vision task of denoising, the purpose is to prepare images that result from physical measurements for further processing that may be the extraction of quantitative information or image segmentation. In order to make this quantitative processing step more stable and reliable, noise and measurement errors should be filtered beforehand.

Linear Diffusion Filtering.   The oldest filtering method for image denoising, linear diffusion filtering, goes back to the works of Iijima and later Gabor, Witkin, and Koenderink [30, 45, 54, 112]. It consists of taking a given (possibly noisy) image u_0 as the initial data to the heat equation

$$\begin{aligned} \partial_t u - \Delta u &= 0 & &\text{on } \mathbb{R}_0^+ \times D, \\ u(0, x) &= u_0(x) & &\text{in } D, \end{aligned} \tag{3.2}$$

with natural boundary conditions (NBC), see Chapter 2.

The PDE results from the minimization of the classical Dirichlet integral

$$E_D[v] := \frac{1}{2} \int_D |\nabla v|^2 \, dx \tag{3.3}$$

by means of a gradient descent in the standard L^2-metric (see Section 2.1) and with u_0 as initial data, i.e.,

$$u = \operatorname*{argmin}_{v \in H^1(D)} E_D[v].$$

Note that in (3.2), we have $u = u(t, x)$, where t is an additional free variable and an artificial parameter that connects with the parabolic nature of the PDE. This parameter acts as a *scale* for filtering. It runs from $t = 0$ with the original image $u(0, x) = u_0(x)$ to coarser and more filtered versions $u(t, x)$ of the original image for increasing t. The linear diffusion filtering (3.2) is an example of a *scale-space* method for which a mathematical framework has been discussed by Alvarez, Guichard, and Lions [3].

However, from the viewpoint of applications, it is obvious that the heat equation leads to a loss of information in the image, as there is no distinction made between noise and data (see Figure 3.2). In fact, with increasing scale t, the initial image gets smoother and smoother (with the limit for $t \to \infty$ being an image with constant image value), which renders the heat equation unsuitable or useful for very short time t only.

The discretization of the heat equation was discussed in Chapter 2.

Figure 3.2: Example for linear and edge-preserving Perona-Malik diffusion. For the given input image (*left*), linear diffusion (*middle*) results in an image with smoothing across object boundaries, whereas Perona-Malik diffusion (*right*) preserves the object boundary.

Nonlinear Diffusion Filtering.   More successful image denoising models have been presented with the selective smoothing of Perona-Malik [80] and Catte et al. [16]. They defined a nonlinear diffusion model with a decreased diffusion coefficient in the vicinity of edges. The resulting evolution problem can be written as

$$\partial_t u - \mathrm{div}\left(g(|\nabla u_\sigma|)\nabla u\right) = f(u) \quad \text{on } \mathbb{R}_0^+ \times D,$$
$$u(0, x) = u_0(x) \quad \text{in } D. \tag{3.4}$$

Here, the diffusion coefficient g acts as a control over the diffusivity. It is designed to yield small values when the gradient of u_σ has a large magnitude, thus indicating the presence of an edge in u_σ. The diffusion operator reduces the diffusion in the vicinity of edges, which thereby improves the behavior of this method vs. the linear filtering from above. An example of a meaningful diffusion coefficient is $g(s) = (1 + s^2/\lambda^2)^{-1}$, where $\lambda > 0$ is a user-prescribed parameter that influences the location of the inflection point of the graph of g.

To properly distinguish noise from edges in the image (and to make the operator mathematically well-posed), a prefiltered version u_σ is used to indicate edges in the diffusion coefficient. Usually one sets u_σ to be the convolution of u with a Gaussian kernel of variance $\sigma > 0$, i.e., $u_\sigma(t, x) = G_\sigma * u(t, x)$. This operation is realized numerically by a short time-step of the linear diffusion (3.2).

Finally, the right-hand-side f (the reaction term) acts as a force that keeps the filtered image u close to the original noisy image u_0, e.g., $f(u(x)) = |u(x) - u_0(x)|^2$. Obviously, if we set $g \equiv 1$ and $f \equiv 0$, we get back to the original linear diffusion filtering (3.2).

Further extensions of the nonlinear diffusion filtering have been obtained by changing the diffusion coefficient and also by replacing it with a diffusion matrix that allows for anisotropic diffusion. See, e.g., [108] for further reading.

The discretization of the nonlinear diffusion filtering can be achieved following the expositions for the heat equation from Chapter 2. To tackle the nonlinearity, a semi-implicit scheme can be used in which the diffusion coefficient and the right-hand side are evaluated at the old time-step.

3.3   RANDOM WALKER AND DIFFUSION IMAGE SEGMENTATION

One of the various fundamental tasks of computer vision and image processing is *segmentation*, in which an image is partitioned into regions that are believed to belong to different objects of interest. Image segmentation is of high importance in many application areas in which the segmented objects are needed for further processing. In medical image processing, for example, organs, tissues, and lesions are segmented from image data to visualize them, to measure their volume, or to take them as computational domains for biophysical simulation. The task of image segmentation will be the topic of the following three sections.

A very simple segmentation method results from a modification of the linear diffusion filtering discussed in the previous section. Let u_0 denote an image that we believe to be segmented into an *object* region and a *background*. For some function w, we consider a weighted variant of the Dirichlet integral (3.3)

$$E_{RW}[v] := \frac{1}{2} \int_D w(x) |\nabla v|^2 \, dx \qquad (3.5)$$

for which we restrict the space of possible solutions by the BC

$$v = 1 \text{ on } V_O, \qquad v = 0 \text{ on } V_B.$$

These Dirichlet BC on the sets V_O and V_B (cf. (DBC), Chapter 2) play the role of a user-prescribed initialization of the object (V_O) and the background (V_B) to be segmented from the given image u_0. We define the seeded domain as $V_S = V_O \cup V_B$ and the unseeded domain as $V_U = D \setminus V_S$.

The necessary condition for a minimum u of the weighted Dirichlet integral (i.e., the Euler-Lagrange equation) (3.5) results in the PDE:

$$\begin{aligned} -\text{div}(w \nabla u) &= 0 \text{ in } D, \\ u &= 1 \text{ on } V_O, \\ u &= 0 \text{ on } V_B. \end{aligned} \qquad (3.6)$$

This is the Poisson equation (EE) with inhomogeneous BC (IDBC) that we have discussed in Chapter 2. The solution of the PDE will be a continuous image function u whose values range between 0 and 1 (by maximum principle), and where the value 1 is attained in some subset of the object and the value 0 is attained in some subset of the background. With a thresholding at, e.g., $1/2$,the image u can finally be partitioned into object $\{x|u(x) \geq 1/2\}$ and background $\{x|u(x) < 1/2\}$. The weight w is supposed to act as an edge indicator function.

In random walker segmentation, a discretization of the PDE (3.6) is considered in the following way: the input image is identified with a graph in which pixels/voxels are the graph nodes, and every pixel/voxel is connected to the neighboring pixels/voxels along the lines of a regular quadrilateral/hexahedral grid by a weighted edge.

If we split off the pixels/voxels that are seeded with BC from the pixels that are actual degrees of freedom by $u_S = u|_{V_S}$ and $u_U = u|_{V_U}$, the discretized version of (3.6) reads

$$Lu_U = -B^T u_S.$$

Here, L and B are matrices that represent the discretized version of the weighted diffusion operator, defined through

$$M_{ij} = \begin{cases} \sum_{\substack{\text{neighboring} \\ \text{nodes } x_k}} w^{ik} & \text{if } i = j, \\ -w^{ij} & \text{if } x_i \text{ and } x_j \text{ are neighboring nodes}, \\ 0 & \text{else}, \end{cases} \tag{3.7}$$

and $L = M|_{(i,j):x_i,x_j \in V_U}$ and $B = M|_{(i,j):x_i \in V_U, x_j \in V_S}$. Thus, L and B are the restrictions of the matrix M that represent the dependence of unseeded nodes on unseeded nodes or unseeded nodes on seeded nodes, respectively.

Finally, the weights w^{ij} are edge indicators on the given image u_0 that are achieved as the exponential of a normalized squared difference, thus

$$w^{ij} = \exp\left(-\beta \frac{\left(u_0^i - u_0^j\right)^2}{\max_{kl}\left(u_0^k - u_0^l\right)^2}\right). \tag{3.8}$$

The parameter β is the only free parameter that has to be chosen by the user. Thus, our discrete weights w^{ij} are actually an approximation of the edge indicator

$$w(x) = \exp\left(-\beta \frac{|\nabla u_0(x)|^2}{\sup_D |\nabla u_0|^2}\right). \tag{3.9}$$

The result of a random walker segmentation of an ultrasound image is shown in Figure 3.3.

The random walker segmentation had originally been introduced as a method simulating random walks on a weighted graph. We focus here on PDEs, thus emphasizing the relation to

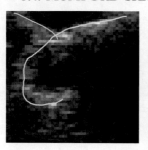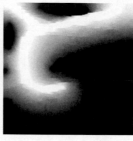

Figure 3.3: (*left*) Definition of the seed regions for the object (V_O, yellow) and the background (V_B, red). (*middle*) The image function u represents the probability that a random walker reaches an object seed. Black denotes probability 0, white probability 1. (*right*) Random walker segmentation result of the ultrasound image. As input, we used the seed regions from the left image and $\beta = 200$.

diffusion and the weighted Dirichlet integral. In its original presentation, the image function u represents the probability for reaching the seeded region V_O from every pixel/voxel. The original description of the random walker method can be found in [36], and an extension and further details are described in [35].

Of course, the random walker segmentation is just an instance of diffusion-based image segmentation. Different choices of edge indicators (here through the weights w^{ij}) can be chosen as a diffusion coefficient or diffusion tensor to realize the partition of an image into segments. A discretization is achieved along the lines of the FDM for the Poisson equation (EE) that we discussed in Chapter 2.

3.4 MUMFORD-SHAH AND AMBROSIO-TORTORELLI SEGMENTATION

The definition of an image processing or computer vision filter as the minimization of an energy functional is a powerful tool, which we discussed above for diffusion filtering. In this spirit, for segmentation of an image u_0, Mumford and Shah [70] proposed the functional

$$E_{MS}[u, K] := \frac{1}{2} \int_{D \setminus K} (u - u_0)^2 dx + \mu \int_{D \setminus K} |\nabla u|^2 dx + \nu \, \mathcal{H}^{d-1}(K) \qquad (3.10)$$

for which an image $u : D \to \mathbb{R}$ and a set of edges $K \subset D$ of this image are sought simultaneously as minimizers. The first term measures the distance between u and the given image u_0, and the second term quantifies the smoothness of u, both away from edges, i.e., on the set $D \setminus K$ only. The third term yields the length of the edge set K through the $(d-1)$D Hausdorff measure of K. The parameters μ and ν are chosen by the user. If the functional is interpreted from the viewpoint of constrained optimization, these user-defined parameters play the role of Lagrange multipliers. A minimizer of (3.10) is characterized by u being "close" to u_0, "smooth" away from edges, and having a "small" set of edges K.

The direct minimization of the Mumford-Shah energy is difficult because of the different nature of u and K: u is a function and K is a set. In addition, the functional is not differentiable, and thus the definition of minimizers based on the Euler-Lagrange equations is impossible. Instead, regularizations have been proposed (see, e.g., [6]). We review the regularization presented by Ambrosio and Tortorelli [4] below.

Ambrosio-Tortorelli Regularization.   To simplify the handling of the edge set K, Ambrosio and Tortorelli replaced it by a *phase field function* $\phi : D \rightarrow \mathbb{R}$. It is a smooth indicator function defined on the whole domain, such that it has value 0 on K and smoothly tends to 1 away from K. The regularized functional is

$$E_{AT}[u, \phi] := E_{\text{fid}}[u] + E_{\text{reg}}[u, \phi] + E_{\text{phase}}[\phi] \,, \tag{3.11}$$

with

$$E_{\text{fid}}[u] = \frac{1}{2} \int_D (u - u_0)^2 \, dx$$

$$E_{\text{reg}}[u, \phi] = \mu \int_D (\phi^2 + k_\varepsilon) |\nabla u|^2 dx \tag{3.12}$$

$$E_{\text{phase}}[\phi] = \nu \int_D \left(\varepsilon |\nabla \phi|^2 + \frac{1}{4\varepsilon} (1 - \phi)^2 \right) dx \,.$$

In this regularized version, the smoothness term has been replaced by an integral E_{reg} that is weighted with the phase field function. Also, the Hausdorff measure has been replaced by the term E_{phase}, which drives the phase field to the value 1 and controls its smoothness. The parameter $\varepsilon > 0$ controls the scale of the edges detected, $\mu > 0$ the amount of detected edges, $\nu > 0$ the behavior of the phase field, and $k_\varepsilon > 0$ ensures positivity of the integral, which translates into coerciveness of the resulting PDE.

Again, a minimizer can be found by transforming (3.12) into the Euler-Lagrange equations of (3.11):

$$-\text{div}\left(\mu(\phi^2 + k_\varepsilon)\nabla u\right) + u = u_0,$$

$$-\varepsilon \Delta \phi + \left(\frac{1}{4\varepsilon} + \frac{\mu}{2\nu} |\nabla u|^2 \right) \phi = \frac{1}{4\varepsilon} \,, \tag{3.13}$$

which is a system of two coupled elliptic PDEs. The BC can be chosen, for example, to be of the Neumann type. The system can be solved by a fixed point iteration in which the first or second equation is solved by keeping the other quantity fixed. An example of Ambrosio-Tortorelli segmentation is shown in Figure 3.4.

Edge Continuity and Consistency.   As can be seen in Figure 3.4, the model (3.13) can lead to partially detected and broken contours. To remedy this problem, Erdem et al. [26] replaced

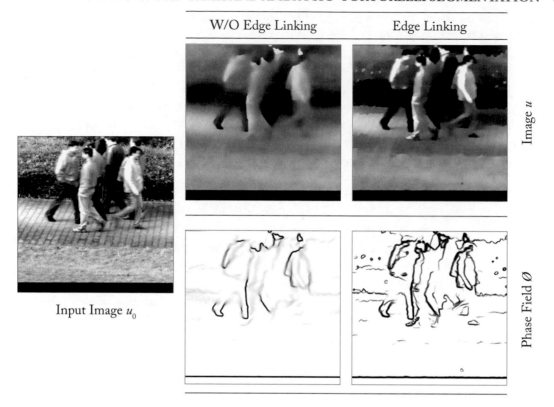

Figure 3.4: Comparison of the Ambrosio-Tortorelli model (*left*) and the extended model using the edge linking procedure (*center and right*). (Input image courtesy of Christoph S. Garbe.)

the diffusion coefficient in the u-equation of (3.13), yielding

$$-\text{div}\Big(\mu\left((c_{dc}c_h\phi)^2 + k_\varepsilon\right)\nabla u\Big) + u = u_0 \,.$$

The product $c_{dc}c_h$ is meant to increase or decrease the diffusivity to smooth away "unwanted" edges or allow for the creation of new edges, respectively.

The directional consistency factor c_{dc} increases diffusion if image gradients are not aligned in the vicinity of detected edges as this would indicate unreliable edges. This effect is achieved by

$$c_{dc} = c_{dc}[u] := \zeta[u] + \frac{1 - \zeta[u]}{\phi},$$

where ζ is the measure of alignment of gradients of the image u, which is defined discretely on all pixels/voxels x_i, for $i \in \mathcal{I}$ by

$$\zeta_i[u] = \exp\left(\varepsilon^{dc}\left(\frac{1}{|\eta_s|}\sum_{j\in\eta_s}\frac{\nabla u_i}{|\nabla u_i|}\cdot\frac{\nabla u_j}{|\nabla u_j|}-1\right)\right). \qquad (3.14)$$

In the expression above, ε^{dc} is another user-prescribed parameter and η_s is a set of s pixels/voxels that lie in a plane that is perpendicular to ∇u_i.

In [26], the continuity of edges is enhanced with the factor $c_h[\phi] = \frac{1}{1+\phi-\phi^2}$, which, however, in our computations we modified to

$$c_h = c_h[\phi] := \frac{1}{1 + \alpha\,(\phi - \phi^2)}$$

for yet another user-prescribed parameter $\alpha > 0$.

The modified Ambrosio-Tortorelli model with edge linking can be solved as before with a fixed point iteration alternating between the two equations. Either the elliptic steady state equations can be solved directly or one can use parabolic versions of the PDEs, thus implementing a gradient descent for the minimization of corresponding energy functionals. Figure 3.4 shows some results obtained from the Ambrosio-Tortorelli segmentation with and without the edge-linking step.

3.5   LEVEL SET METHODS FOR IMAGE SEGMENTATION

A different viewpoint on image segmentation is reached, in contrast to the Mumford-Shah/Ambrosio-Tortorelli segmentation, when dealing with the evolution of curves. A curve that is initialized by the user is thereby driven toward the boundaries of an object that is to be segmented.

To represent such a curve numerically, both an explicit [51] and an implicit treatment [14, 18, 21, 63, 94] have been presented in the literature. Explicit representations, however, have some drawbacks that make their application complicated: parameterization of the curve, distribution of the nodes describing the curve, dependence of the result on the parameterization, etc.

Therefore, implicit methods have become increasingly popular, especially the *level set method* in the image processing and computer vision community. The method was originally derived independently by Dervieux and Thomasset [21] and Osher and Sethian [75, 94] and is closely related to the phase field method that we briefly touched on in the previous section in the context of the Ambrosio-Tortorelli segmentation.

Central to the idea of the level set method is to identify a curve $C_0 \subset \mathbb{R}^d$ with the zero level set of a function $\phi : \mathbb{R}_0^+ \times \mathbb{R}^d \to \mathbb{R}$, $(t, x) \mapsto \phi(t, x)$ such that $C_0 = \{x \in \mathbb{R}^d : \phi(0, x) = 0\}$. An evolution of the curve yielding

$$C(t) = \{x \in \mathbb{R}^d : \phi(t, x) = 0\} \qquad \text{for all } t > 0$$

is then obtained by changes of the function ϕ according to

$$\partial_t \phi + F|\nabla\phi| = 0, \tag{3.15}$$

where $F : \mathbb{R}_0^+ \times \mathbb{R}^d \to \mathbb{R}$ is the speed of the curve in its normal direction. An initial value for ϕ is chosen appropriately, either automatically or by user interaction.

We now switch the type of PDEs from parabolic or elliptic, i.e., the diffusion-dominant second-order equations that we discussed in previous sections, to hyperbolic, i.e., the transport-dominant first-order equations. For the numerical treatment of (3.15), Osher and Sethian [75] developed numerical methods based on upwinding known from Hamilton-Jacobi equations. In addition, Sethian [94] developed efficient methods, such as the Narrow Band Method, in which (3.15) is solved in the vicinity of the zero level set only.

For various reasons, it is desirable to let the level set function ϕ be a signed distance function, i.e., $|\nabla\phi| = 1$. This property, however, is not maintained during the evolution. For recovery of this property, reinitializations are necessary from which the level set method is started again. Such reinitialization can be realized through solving $\partial_t \phi = \text{sign}(\phi)(1 - |\nabla\phi|)$ until a steady state is reached or, more efficiently, with the Fast Marching Method (FMM) [94].

For the curve evolution models discussed in the remainder of this section, the design of the speed F and additional regularizations will bring back the elliptic/parabolic character of the equations, such that they can be treated numerically with the FDM or FEM that we reviewed in Chapter 2.

Gradient-Based Segmentation. A straightforward approach to the curve evolution is to use the gradient field of a given image u_0 to drive a user-initialized curve toward steep gradients that indicate an object's boundary. In this spirit, Caselles et al. [14] proposed to use

$$F = g[u_0](1 - \varepsilon\kappa) \quad \text{with} \quad g[v] := \frac{1}{1 + |\nabla G_\sigma * v|^2} \tag{3.16}$$

in (3.15). This force scales down the curve evolution in the vicinity of edges and increases the speed in smooth regions. Moreover, a curvature smoothing is obtained with the user-prescribed parameter ε that controls the influence of the mean curvature κ of the level set function. Less smoothing is obtained for high curvature of the level set.

With a suitable stopping criterion, this method achieves quite good results in the presence of strong gradients, an example of which is shown in Figure 3.5.

Geodesic Active Contours. Caselles et al. [15] and, simultaneously, Kichenassamy et al. [53] developed geodesic active contours. They start from an energy that is formulated for explicit curves that measure smoothness and "external" energy resulting from edges of objects in the image u_0. Formulated as a level set equation, the geodesic active contour model reads

$$\partial_t \phi = -\alpha\nabla g[u_0] \cdot \nabla\phi - \beta g[u_0]|\nabla\phi| + \varepsilon\kappa|\nabla\phi|, \tag{3.17}$$

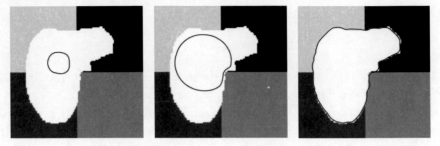

Figure 3.5: Segmentation of a sample image based on a level set propagation with gradient-based speed function. The time/scale increases from left to right and the zero level set (red line) approximate the boundary of the object (a liver mask) at the end.

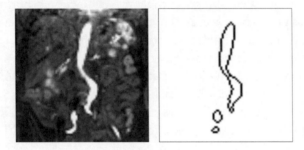

Figure 3.6: Segmentation using geodesic active contours. (*left*) The initial image. (*right*) Solution of the geodesic active contour method initialized with small circles inside the object.

where $g[u_0]$ is the force defined above in (3.16) and where α, β, and ε are user-defined parameters.

In contrast to the plain gradient-based segmentation, this is a level set equation that contains an additional advection term with velocity $\alpha \nabla g[u_0]$, which forces the zero level set to stay in regions with high gradients. Figure 3.6 shows a typical geodesic active contours segmentation result.

Chan-Vese Segmentation. The previous segmentation methods all relied on steep gradients as the indicators for edges of objects. When such a gradient is not present, the methods perform poorly or even fail. Chan and Vese [18] proposed a method that is independent of gradient information. Instead, their model relies on homogeneous regions inside the object and the background, which is achieved by searching for image values $c_1, c_2 \in \mathbb{R}$ as well as a curve $C \subset D$ that minimize the functional

$$
\begin{aligned}
E_{CV}[c_1, c_2, C] = {} & \mu \, \mathcal{H}^{d-1}(C) + \nu \, \mathcal{H}^d \, (\text{inside}(C)) \\
& + \lambda_1 \int_{\text{inside}(C)} |u_0 - c_1|^2 \, dx + \lambda_2 \int_{\text{outside}(C)} |u_0 - c_2|^2 \, dx
\end{aligned}
\tag{3.18}
$$

for user-chosen parameters $\mu, \nu, \lambda_{1,2}$. This functional measures the length of the curve and the area enclosed by it through the $(d-1)$-dimensional and d-dimensional Hausdorff measures of C and inside(C), respectively. The fidelity terms measure the distance between the given image u_0 and the average gray values $c_{1,2}$ in the inside and outside, i.e., in the object and the background.

Again, for ease of handling, it is convenient to express C as the zero level set of a signed distance function ϕ. Then the minimization of (3.18) can be done in an alternating fashion. First, it must be noted that for fixed ϕ, the optimal values of $c_{1,2}$ can be computed directly as

$$c_1[\phi] = \frac{\int_D u_0(x) H_\varepsilon(\phi(x))\, dx}{\int_D H_\varepsilon(\phi(x))\, dx} \quad \text{and} \quad c_2[\phi] = \frac{\int_D u_0(x)\,(1 - H_\varepsilon(\phi(x)))\, dx}{\int_D (1 - H_\varepsilon(\phi(x)))\, dx} \tag{3.19}$$

with a regularized Heaviside function H_ε that allows integrating inside and outside the contour:

$$H_\varepsilon(z) = \frac{1}{2}\left(1 + \frac{2}{\pi}\arctan\left(\frac{z}{\varepsilon}\right)\right).$$

The derivative of H_ε, a regularized Dirac distribution δ_ε, allows writing a gradient descent to reach the Euler-Lagrange equation that results from the necessary condition for a minimum of E_{CV} in the level set function $\phi = \phi[c_1, c_2]$ for fixed $c_{1,2}$:

$$\partial_t \phi = \delta_\varepsilon(\phi)\left(\mu \operatorname{div}\left(\frac{\nabla\phi}{|\nabla\phi|}\right) - \nu - \lambda_1(u_0 - c_1)^2 + \lambda_2(u_0 - c_2)^2\right). \tag{3.20}$$

An initial value and BCs can be chosen appropriately. This equation is not of the classical form of hyperbolic level set equation (3.15) but rather a parabolic PDE that contains a curvature smoothing term, a term penalizing the segmented area, and two terms penalizing variations from the mean values of the segmented object (c_1) and the background (c_2).

The minimization of (3.18) then lies in the alternating solution of (3.19) and (3.20). Figure 3.7 illustrates the advantage of the Chan-Vese model over gradient-based segmentation, as it does not need a steep image gradient to separate regions. It instead separates regions that have minimal distance to their average image values.

Phase Field Models and Parabolic Regularizations.   Phase field models, such as the one we reviewed for the Ambrosio-Tortorelli segmentation in Section 3.4, are closely related to level set methods. In the level set approach, a curve, e.g., the boundary curve of an object, is represented "sharply" as the zero level set of a function. In the literature, this approach is also referred to as the *sharp interface approach*.

In contrast to level set approaches, the phase field models represent a *diffusive interface approach*. The phase field function is constant away from the interface and varies smoothly across the interface. Thus, phase field models can be regarded as treating an interface that separates two regions with a transition zone that contains mixed content from both sides of the interface.

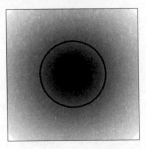

Figure 3.7: Segmentation of an object without sharp edges using the Chan-Vese approach. In red, we show the steady-state solution of the Chan-Vese segmentation method that has been initialized with a small circle inside the object.

The advantage of phase field models over level set methods is that the resulting PDEs are in most cases parabolic instead of hyperbolic and thus much easier to treat numerically. In many cases, it is possible to use a phase field model instead of an analog level set model. This transition either uses second-order operators already built into the level set equation or encompasses explicit parabolic regularization.

It is central to this approach that the mean curvature κ, which is used in many level set equations, can be expressed in various ways. In fact, if the phase field results from a double-well potential, it will behave locally, in normal direction to an edge, like a hyperbolic tangent function. In this case, the mean curvature can be expressed as

$$\kappa := \operatorname{div}\left(\frac{\nabla\phi}{|\nabla\phi|}\right) = \frac{1}{|\nabla\phi|}\left(\Delta\phi + \frac{\phi(1-\phi^2)}{\varepsilon^2}\right) \tag{3.21}$$

in which ε is the "thickness" parameter of the phase field. This formulation of the mean curvature is the projection of the Laplacian on the space tangent to the level set [99]. Consequently, level set equations containing curvature terms can be written as parabolic equations. In this spirit, a phase field version of the gradient-based segmentation (3.15), (3.16) is

$$\partial_t\phi + g[u_0]\,|\nabla\phi| = \varepsilon\, g[u_0]\left(\Delta\phi + \frac{\phi(1-\phi^2)}{\varepsilon^2}\right).$$

In general, a hyperbolic (level set) equation of the form $\partial_t\phi + F|\nabla\phi| + \mathbf{w}\cdot\nabla\phi = 0$ with force F and advection velocity $\mathbf{w}$ (like (3.17)) can be regularized with a curvature smoothing term and interpreted as a phase field equation to yield

$$\partial_t\phi + F|\nabla\phi| + \mathbf{w}\cdot\nabla\phi = b\left(\Delta\phi + \frac{\phi(1-\phi^2)}{\varepsilon^2}\right)$$

for some $b > 0$. If the curvature smoothing is not wanted, i.e., cases in which b should be set to zero, one can keep the parabolic nature of the equation but cancel out the curvature smoothing with a counter term [28] that basically subtracts the curvature from the right-hand side for any $b > 0$, cf. (3.21),

$$\partial_t \phi + F|\nabla\phi| + \mathbf{w}\cdot\nabla\phi = b\left(\Delta\phi + \frac{\phi(1-\phi^2)}{\varepsilon^2} - |\nabla\phi|\kappa\right)$$
$$= b\left(\Delta\phi + \frac{\phi(1-\phi^2)}{\varepsilon^2} - |\nabla\phi|\operatorname{div}\left(\frac{\nabla\phi}{|\nabla\phi|}\right)\right). \tag{3.22}$$

It is important to understand that (3.22) is different from simply setting $b = 0$ because it relies on the assumption that ϕ has the hyperbolic tangent profile in the vicinity of its zero crossing, which was the basis for (3.21).

Departing from this assumption, it is also possible to assume profiles of the phase field function that are different from the tanh-shape. In fact, with a different source term instead of the double-well potential with the term $\phi(1-\phi^2)$, it is possible to obtain the profile of a signed distance function. An example of a different source term has been achieved by Glasner [33] with a *nonlinear preconditioning* of the form

$$\phi_t + F|\nabla\phi| + \mathbf{w}\cdot\nabla\phi = b\left(\Delta\phi + \frac{1}{\varepsilon}\left(1-|\nabla\phi|^2\right)\sqrt{2}\tanh\left(\frac{\phi}{\sqrt{2}\varepsilon}\right) - |\nabla\phi|\operatorname{div}\left(\frac{\nabla\phi}{|\nabla\phi|}\right)\right).$$

In this equation, the first two terms of the right-hand side represent the mean curvature in terms of the signed distance function, whereas the counter-term, i.e., the third term of the right-hand side, is left unchanged.

We will utilize the connection between level set and phase field methods (as well as the regularization and preconditioning) intensively in Chapter 6 in a discussion of stochastic variants of the well-known level set segmentation methods.

3.6 VARIATIONAL METHODS FOR OPTICAL FLOW

We now turn to another important task of computer vision, the optical flow problem in which the apparent motion of objects in an image sequence is analyzed. The goal of this task is to find a vector field, the so-called optical flow field $\mathbf{w} : Q \to \mathbb{R}^d$, that describes this motion along apparent trajectories $x(s)$ such that $\partial_s x(s) = \mathbf{w}(s, x(s))$.

The fundamental basis for most optical flow models is the so-called *brightness constancy constraint* (BCC) assumption that states that the image value of an object does not change along its apparent motion trajectory in an image sequence, thus $u(s, x(s)) = \text{const}$. Differentiating with respect to time yields

$$\partial_s u(s, x(s)) + \mathbf{w}(s, x(s))\cdot\nabla u(s, x(s)) = 0 \qquad \text{for all } s \in [0, T], \tag{3.23}$$

where ∇ denotes the gradient w.r.t. space coordinates x. This is a scalar equation in the $d > 1$ components of $\mathbf{w}$. Thus, the system is underdetermined, i.e., the direct solution of (3.23) is not possible, a fact known as the *aperture problem*. To remedy this problem, regularizations of the optical flow field are needed such that the BCC equation plays the role of a data term only.

Horn and Schunck Model.   The standard Horn and Schunck model [44] deals with the aperture problem by looking for harmonic flow fields, thus minimizing the energy

$$E_{HS}[\mathbf{w}] = \frac{1}{2}\int_D |\partial_s u + \mathbf{w} \cdot \nabla u|^2 \, ds \, dx + \frac{\alpha}{2}\int_D |\nabla \mathbf{w}|^2 \, ds \, dx \tag{3.24}$$

for a given image sequence u and for some positive regularization parameter α. The minimization is achieved by solving its Euler-Lagrange equation. Denoting the coordinates with $x = (x_1, \ldots, x_d)$ for a d-dimensional image sequence, and the components of the optical flow field with $\mathbf{w} = (w_1, \ldots, w_d)$, the PDEs are

$$\partial_1 u \, (\mathbf{w} \cdot \nabla u + \partial_s u) - \alpha \Delta w_1 = 0,$$
$$\vdots \qquad \vdots \tag{3.25}$$
$$\partial_d u \, (\mathbf{w} \cdot \nabla u + \partial_s u) - \alpha \Delta w_d = 0,$$

where $\partial_{1,\ldots,d}$ denotes the partial derivative w.r.t. the respective spatial coordinate $x_{1,\ldots,d}$. Boundary conditions are again chosen appropriately. This is a simple linear system for the unknowns $w_1, \ldots, w_d$, which can be solved straightforwardly. Figure 3.8 shows an application of this optical flow model.

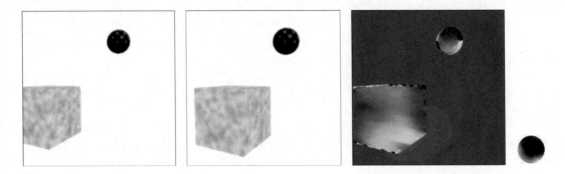

Figure 3.8: The first and last frame of the input sequence (*left and middle*) and the Horn-Schunck optical flow field of the underlying image sequence (*right*). The colorwheel at the right indicates the directions of the flow field.

Discontinuity-Preserving Optical Flow.   The Horn and Schunck energy consists of the data term that enforces the BCC assumption and the regularization, which corresponds to linear and isotropic smoothing of the optical flow field. The drawback of this approach is similar to the drawback of the heat equation for denoising, that the optical flow field is smoothed out over edges and boundaries of objects.

An improvement introduces an edge indicator function as a scaling of the diffusion analog to the step from linear denoising to selective smoothing, which yields the PDE system

$$\partial_1 u \, (\mathbf{w} \cdot \nabla u + \partial_s u) - \alpha \mathrm{div}(g(|\nabla u_\sigma|)) \nabla w_1) = 0,$$
$$\vdots \qquad \vdots \qquad\qquad\qquad (3.26)$$
$$\partial_d u \, (\mathbf{w} \cdot \nabla u + \partial_s u) - \alpha \mathrm{div}(g(|\nabla u_\sigma|)) \nabla w_d) = 0.$$

Combined Local Global Method.   A different regularization approach comes from smoothing the data term. In the combined local global method presented by Bruhn and Weickert [10], trajectories in the sequence are characterized through local coherence as in the approach from Lucas and Kanade [61]. Thus, a component-wise smoothed time-space structure tensor

$$J_\sigma(s,x) := \left(\begin{pmatrix} \partial_s u \\ \nabla u \end{pmatrix} (\partial_s u, \nabla u) \right)_\sigma \in \mathbb{R}^{(d+1) \times (d+1)}$$

for some $\sigma > 0$ is considered in the data term. Furthermore, Bruhn and Weickert replaced the Horn and Schunck regularizer by a time-space gradient, yielding the energy

$$E_{CLG}[\mathbf{w}] := \frac{1}{2} \int_Q \left| \begin{pmatrix} 1 \\ \mathbf{w} \end{pmatrix} \cdot J_\sigma \begin{pmatrix} 1 \\ \mathbf{w} \end{pmatrix} \right|^2 ds\, dx + \frac{\alpha}{2} \int_Q \left| \begin{pmatrix} \partial_s \mathbf{w} \\ \nabla \mathbf{w} \end{pmatrix} \right|^2 ds\, dx. \qquad (3.27)$$

As before, the minimization yields a linear system of d PDEs of the form

$$(J_\sigma)_{k2} w_1 + \cdots + (J_\sigma)_{k(d+1)} w_d + \alpha(\partial_{ss} + \Delta) w_k = -(J_\sigma)_{k1} \qquad (3.28)$$

for $k = 1, \ldots, d$.

3.7   VARIATIONAL METHODS FOR REGISTRATION

The final application that we consider here is the matching of two different images, known as image registration. A deformation vector field $\mathbf{w} : D \to \mathbb{R}^d$ is sought that describes a correspondence between pixels/voxels in the one image, called reference image u, and another image, called template image v, and such that an energy becomes minimal.

Analog to the optical flow problem, image registration is also ill-posed. Therefore, energies consist of data terms and regularization terms, which constrain the set of possible deformation vector fields. Among simple smoothness regularizers, there have been investigations on physically motivated regularizers, as in fluid registration or elastic registration.

In elastic registration, the regularizer is motivated by considering the objects visible in images to be elastic deformable bodies. Then, the energy contains a standard fidelity term and the linear elastic potential of $\mathbf{w}$ as a regularizer. Denoting the deformed template image with $v_{\mathbf{w}} := v \circ (\mathrm{Id} - \mathbf{w})$, we have

$$E_{ER}[\mathbf{w}] := \frac{1}{2} \int_D |v_{\mathbf{w}} - u|^2 \, dx + \frac{\mu}{4} \int_D \left| \nabla \mathbf{w} + (\nabla \mathbf{w})^T \right|^2 \, dx + \frac{\lambda}{4} \int_D |\mathrm{div}(\mathbf{w})|^2 \, dx. \qquad (3.29)$$

The parameters $\lambda, \mu > 0$ are known as the *Navier–Lamé constants*.

Again, minimization of this functional leads to a linear system of PDEs

$$-\mu\Delta\mathbf{w} - (\lambda + \mu)\nabla\text{div}(\mathbf{w}) = (v_{\mathbf{w}} - u)\nabla v_{\mathbf{w}} \tag{3.30}$$

together with meaningful boundary values, like homogeneous Dirichlet BC (DBC) for $\mathbf{w}$. An example for linear elastic registration is shown in Figure 3.9. For further details, we refer the reader to [68] and the references therein.

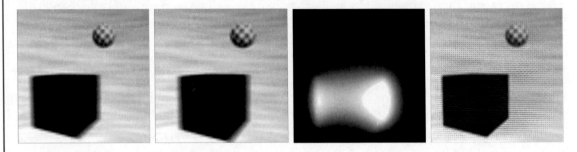

Figure 3.9: Exemplary reference (*left*) and template (*middle left*) for elastic registration. The resulting vector field component in the x-direction (*middle right*) and deformation field (*right*).

Note 3.7   Readers interested in advanced usage of PDE-based image processing may consult the following books:

- [69]: Sophisticated approaches for image registration

- [94]: Detailed introduction into level sets with applications in image processing

- [6]: Advanced variational problems in image processing

CHAPTER 4

Numerics of Stochastic PDEs

Note 4.1   Readers not familiar with the numerics of stochastic PDEs may consult the following literature:

- [115]: An excellent overview written by the polynomial chaos pioneer

- [62]: Overview of spectral methods for numerical uncertainty quantification

This chapter deals with the fundamentals required to develop stochastic images. First, we review notation and results from probability theory. We then introduce stochastic PDEs (SPDEs) and the polynomial chaos expansion, the main ingredient we will use for the numerical approximation of random variables. The discretization of SPDEs is an active research field. Besides the discretization based on sampling approaches such as Monte Carlo simulation or stochastic collocation, methods for the *intrusive computation* of stochastic solutions have been discussed in the literature. "Intrusive" means that we do not generate the solutions based on sampling strategies and we cannot reuse deterministic algorithms to compute these solutions at sampling points. Instead, we need to develop new algorithms that are more efficient than the classical sampling approaches.

This book focuses on intrusive methods. We present sampling-based approaches as well, but we use them to verify the correctness of the intrusive algorithms and implementations only. The intrusive methods presented herein range from the stochastic FDM and the stochastic FEM to the generalized spectral decomposition, which is a method that allows speeding up of the solution process of the stochastic FEM (SFEM); see Figure 4.1.

4.1   BASICS FROM PROBABILITY THEORY

This section provides background from probability theory for the presentation of the concept of stochastic images and SPDEs. First, we introduce the basic ingredients, probability measures, probability spaces, and random variables.

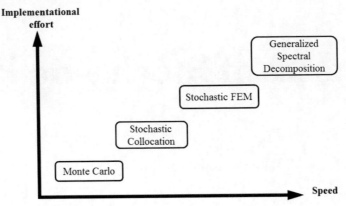

Figure 4.1: Comparison of discretization methods in terms of the implementation effort and speed of convergence toward the true solution.

Definition 4.2   A *probability space* $(\Omega, \mathcal{A}, \Pi)$ is a triplet consisting of a sample space Ω containing all possible outcomes, a σ-algebra of events $\mathcal{A} \subset 2^{\Omega}$ (i.e., a subset of the power set of Ω), and a probability measure Π. The *probability measure* Π is defined on the σ-algebra $\mathcal{A}$ and has the following properties:

- Π is nonnegative: $\Pi(A) \geq 0$ for all $A \in \mathcal{A}$.

- The measure of the sample space Ω is one: $\Pi(\Omega) = 1$.

- Π is countable additive, i.e., for a countable number of pairwise disjoint sets $A_i \subset \mathcal{A}$, we have $\Pi(\cup A_i) = \sum_i (\Pi(A_i))$.

On the probability space $(\Omega, \mathcal{A}, \Pi)$, we define functions into the real numbers.

Definition 4.3   A real-valued *random variable* $f : \Omega \to \mathbb{R}$ is a function from the sample space Ω into the real numbers. It must be measurable with respect to the σ-algebras $\mathcal{A}$ and $\mathcal{B}$, where $\mathcal{B}(\mathbb{R})$ is the Borel σ-algebra on $\mathbb{R}$.

Random variables are important descriptors for the definition of stochastic images. In Chapter 5, we will see that every pixel of a stochastic image can be considered to depend on random variables. For random variables, it is possible to define the probability density function:

Definition 4.4   The function $\rho : \mathbb{R} \to \mathbb{R}$ is called a *probability density function* (PDF) of a continuous random variable f if it satisfies $\Pi(a < f < b) = \int_a^b \rho(x)\,dx$ for all $a, b \in \mathbb{R}$.

Having the probability density at hand, we define further properties of random variables, the most important of which is the expected value.

Definition 4.5   The *expected value* or first moment of a random variable $X : \Omega \to \mathbb{R}$ with PDF ρ is

$$\mathbb{E}(X) = \int_\Omega X(\omega)\,d\Pi = \int_{\mathbb{R}} x\rho(x)\,dx = \int_{\mathbb{R}} x\,d\Pi_X. \tag{4.1}$$

In (4.1) we used the induced measure $d\Pi_X = \rho\,dx$ to characterize integration with respect to the PDF. Thus, knowing the probability density of a random variable allows us to transform the integral over the sample space Ω into an easier computable integral over the real numbers weighted by the probability density. Using this equality, it is also possible to compute higher-order moments of random variables.

Definition 4.6   The nth *central moment* of a random variable X is

$$M^n(X) = \mathbb{E}((X - \mathbb{E}(X))^n). \tag{4.2}$$

A useful member of this class of moments is the second central moment, the *variance*:

$$\text{Var}(X) = \mathbb{E}\left((X - \mathbb{E}(X))^2\right). \tag{4.3}$$

To evaluate the relation between different random variables, we use the covariance.

Definition 4.7   The *covariance* of two random variables f, g with finite second-order moments is

$$\text{Cov}(f, g) = \mathbb{E}(fg) - \mathbb{E}(f)\mathbb{E}(g). \tag{4.4}$$

In what follows, it will be necessary to have a set of random variables indexed by a spatial position. Random fields are elements of a tensor product space consisting of functions defined on the Cartesian product $\Omega \times D$. A random field is a function taking two arguments: a random outcome and a spatial position. We restrict the investigations to random fields satisfying smoothness assumptions.

Definition 4.8 Let $D \subset \mathbb{R}^d$ be a spatial domain and Ω the sample space. For some Banach space $Y(D)$, we define the tensor space $L^2(\Omega) \otimes Y(D)$ of *random fields* as the space of random variables u satisfying $u(\omega, \cdot) \in Y(D)$ almost surely and $u(\cdot, x) \in L^2(\Omega)$ for all $x \in D$, and

$$L^2(\Omega) = \left\{ f : \Omega \to \mathbb{R} : \mathbb{E}(f^2) < \infty \right\} . \tag{4.5}$$

The phrase "almost surely" (or "a.s.") used here, and in the remainder of the section, denotes that an event happens with probability 1. Thus, in the above definition, "$u(\omega, \cdot) \in Y(D)$ almost surely" means that $\Pi(u(\omega, \cdot) \in Y(D)) = 1$.

Of course, for the stochastic component of random fields, we utilize the definition of expected value, moments, and variance as defined above. Additionally, on the space $L^2(\Omega)$, we also define the inner product/scalar product

$$\langle f, g \rangle = \int_{\Omega} f(\omega)g(\omega) \, d\Pi,$$

which makes $L^2(\Omega)$ a Hilbert space. We also write $\langle f^2 \rangle = \langle f, f \rangle$. Note that $\langle f^2 \rangle = \mathbb{E}(f^2)$.

The finiteness of the second-order moment claimed in (4.5) is a strong limitation that is typically not satisfied for random fields arising in financial problems [43, 71]. However, this assumption seems reasonable in our scope due to the limited energy an image acquisition device detects. Concerning the Banach space Y for image processing problems, we will work with the Hilbert spaces $L^2(D)$ and $H^1(D)$, as those are the regularities typically assumed for classical image processing tasks [6]. Thus, we will be dealing with random fields from $L^2(\Omega) \otimes L^2(D)$ and $L^2(\Omega) \otimes H^1(D)$.

4.2 STOCHASTIC PARTIAL DIFFERENTIAL EQUATIONS

We introduce SPDEs following [7] and by using an elliptic model equation analog to our discussions from Chapter 2. Let $D \subset \mathbb{R}^d$ be a bounded domain on which we consider the deterministic PDE

$$\begin{aligned} -\mathrm{div}(w\nabla u) &= f \quad \text{on } D \\ u &= g \quad \text{on } \partial D . \end{aligned} \tag{4.6}$$

Here, w is a diffusion coefficient, f a source term, and g the boundary condition (cf. (EE), Chapter 2). In this equation, we assume that we perfectly know the diffusion coefficient w, the right-hand side f, and the domain D. Therefore, we call this formulation deterministic.

The motivation to transform this problem into its stochastic variant is (cf. Chapter 1) that in many applications, these quantities are not known exactly. Instead, a description of the quantities through random fields is possible (see e.g., [2]). Now, consider $a : \Omega \times \bar{D} \to \mathbb{R}$ a stochastic function with continuous and bounded covariance function that satisfies

$$\exists w_{min}, w_{max} \in (0, \infty) \quad \text{with} \quad \Pi \left(\{ \omega \in \Omega \mid w(\omega, x) \in [w_{min}, w_{max}], \forall x \in \bar{D} \} \right) = 1. \quad (4.7)$$

Therefore, we have a stochastic diffusion coefficient that is bounded away from zero and infinity for realizations $\omega \in \Omega$ almost surely. In addition, let $f : \Omega \times \bar{D} \to \mathbb{R}$ be a stochastic function that satisfies

$$\int_{\Omega} \int_{D} f^2(\omega, x) \, dx \, d\Pi = \mathbb{E} \left(\int_{D} f^2(\omega, x) \, dx \right) < \infty, \quad (4.8)$$

i.e., $\mathbb{E}(\|f\|_{2,D}^2) < \infty$. Then the elliptic SPDE analog to 4.6 reads

$$\begin{aligned} -\mathrm{div}(w(\omega, x) \nabla u(\omega, x)) &= f(\omega, x) & \text{almost surely on } D, \\ u(\omega, x) &= g(x) & \text{almost surely on } \partial D, \end{aligned} \quad (4.9)$$

for which a solution $u \in L^2(\Omega) \otimes H^1(D)$ is sought. A proof of the existence and uniqueness of solutions for elliptic SPDEs is closely related to the existence and uniqueness proof of the classical problem, and we refer the reader to [7] for details.

Applying this concept analogously to parabolic and hyperbolic PDEs yields parabolic and hyperbolic SPDEs. For parabolic equations, we have to add time-dependence for the solution and incorporate an additional time-derivative. We end up with a prototype for parabolic SPDEs given by

$$\begin{aligned} \partial_t u(\omega, t, x) - \mathrm{div}(w(\omega, t, x) \nabla u(\omega, t, x)) &= f(\omega, t, x) & \text{almost surely on } (0, T) \times D, \\ u(\omega, t, x) &= 0 & \text{almost surely on } (0, T) \times \partial D, \\ u(\omega, 0, x) &= u_0(x) & \text{almost surely on } D. \end{aligned} \quad (4.10)$$

Vage [105] proved the existence and uniqueness of solutions for this kind of parabolic SPDE. Note that in the spirit of Definition 4.8, we also have random fields for the time-space cylinder $Q = (0, T) \times D$ in $L^2(\Omega) \otimes Y(Q)$, where $Y(Q)$ is a corresponding Banach space on Q. These random fields are then functions $u(\omega, t, x)$ such that $u(\cdot, t, x) \in L^2(\Omega)$ for $(t, x) \in Q$, and further such that $u(\omega, \cdot, \cdot) \in Y(Q)$ almost surely.

For further details on the measure-theoretic and functional-analytic basis behind SPDEs, we refer the reader to [105].

4.3   POLYNOMIAL CHAOS EXPANSIONS

The main foundation for the numerical treatment of SPDEs is the polynomial chaos expansion of random variables. Based on the fundamental work of Wiener [111], who developed the polynomial chaos for Gaussian processes, Cameron and Martin [12] proved that every random variable with a finite variance has a representation as a Fourier-Hermite series. Later, Xiu and Karniadakis [116] developed the Wiener-Askey polynomial chaos or *generalized polynomial chaos* (gPC), which allows a representation of any random process with finite second-order moments in the polynomial chaos with an optimal basis.

The advantage of representing random variables in the generalized polynomial chaos is the simplification of the calculation of integrals over the stochastic part as they are encountered in the computation of moments (4.1)—(4.3). In fact, for arbitrary random variables with an unknown probability density function, we have to calculate the integral over an event space Ω. This event space, however, is abstract, and, depending on the application, not much is known about it.

Now, the use of polynomial chaos expansions allows us to transform the moment-integrals over Ω into an integral over the real numbers by using the probability density function of the underlying random variables. Figure 4.2 shows the situation: instead of the direct computation of the integrals with the random variable X and the measure $d\Pi$, we transform the integral into integration over the real numbers by using the polynomial chaos and the PDF ρ of the underlying random variables. We further elaborate below.

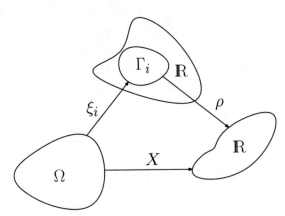

Figure 4.2: Relation between the stochastic spaces. We avoid the integration over Ω with respect to the measure $d\Pi$. Instead, we transform the integral over Ω into integration over a subset of $\mathbb{R}$ (the space Γ_i) with respect to the known PDF ρ_{ξ_i} of the basic random variables ξ_i.

Wiener Chaos.   In his seminal paper [111], Wiener developed the *homogeneous chaos* (or *Wiener chaos*). He formulated the homogenous chaos using Hermite polynomials in independent Gaussian random variables with zero mean and unit variance. Let $\boldsymbol{\xi} = (\xi_1, \dots)$ be a vector of independent Gaussian random variables on Ω with zero mean, unit variance, and PDFs ρ_{ξ_i}. Furthermore, $V_n(\xi_{i_1}, \dots, \xi_{i_n})$ shall be Hermite polynomials in n of these random variables.

Cameron and Martin [12] proved that a random variable X with finite second-order moment has the representation

$$X(\omega) = a_0 V_0 + \sum_{i_1=1}^{\infty} a_{i_1} V_1(\xi_{i_1}(\omega)) + \sum_{i_1=1}^{\infty} \sum_{i_2=1}^{\infty} a_{i_1 i_2} V_2(\xi_{i_1}(\omega), \xi_{i_2}(\omega)) + \dots$$

for any $\omega \in \Omega$. For convenience, this expression can be rewritten using the multi-index notation

$$X(\omega) = \sum_{|\alpha|=0}^{\infty} a_\alpha \Psi^\alpha(\boldsymbol{\xi}(\omega)), \tag{4.11}$$

for $\alpha = (\alpha_1, \alpha_2, \dots)$ and with the obvious correspondence between V_n and Ψ^α. The so-called *Fourier-Hermite* coefficients a_α, sometimes also called *modes*, of the expansion (4.11) are obtained by Galerkin (L^2)-projection, i.e.,

$$a_\alpha = \int_\Omega X(\omega) \Psi^\alpha(\boldsymbol{\xi}(\omega)) \, d\Pi. \tag{4.12}$$

The Hermite polynomials Ψ^α form an orthogonal basis of the space $L^2(\Omega)$, i.e.,

$$\int_\Omega \Psi^\alpha(\boldsymbol{\xi}(\omega)) \Psi^\beta(\boldsymbol{\xi}(\omega)) \, d\Pi = \langle \Psi^\alpha, \Psi^\beta \rangle = \langle (\Psi^\alpha)^2 \rangle \delta_{\alpha\beta}. \tag{4.13}$$

We consider a *finite number* of basic random variables $\xi = (\xi_1, \dots, \xi_n)$ and thus simplify (4.13) by using (4.1) to yield

$$\langle \Psi^\alpha(\xi), \Psi^\beta(\xi) \rangle = \int_\Omega \Psi^\alpha(\boldsymbol{\xi}(\omega)) \Psi^\beta(\boldsymbol{\xi}(\omega)) \, d\Pi = \int_\Gamma \Psi^\alpha(\xi) \Psi^\beta(\xi) \, d\Pi_X, \tag{4.14}$$

where $\Gamma = \mathrm{supp}(\xi) \subset \mathbb{R}^n$ and

$$d\Pi_X = \prod_{i=1}^{n} d\Pi_{\xi_i}, \quad \text{with } d\Pi_{\xi_i} := \rho_{\xi_i} \, dx. \tag{4.15}$$

From the viewpoint of the Hermite polynomials, a weighting function w that is needed to get them orthonormal would be

$$\rho(\xi) = \frac{1}{(2\pi)^{n/2}} e^{-\frac{1}{2}\xi \cdot \xi}.$$

This weighting function is key to understanding the good approximation quality of the Hermite expansion, because $\rho(\xi)$ is the same as the PDF of an n-dimensional Gaussian random variable, i.e., $\rho(\xi) = \prod_i \rho_{\xi_i}$ and thus $d\Pi_X = \rho(\xi)\,dx$. Xiu and Karniadakis [116] further investigated this correspondence between the weighting functions for the orthogonal polynomial basis and the density functions of random variables in the case of other distributions, as we will see in the next section.

In summary, the computation of the scalar product (4.14) reduces to integration over a subset of $\mathbb{R}^n$ with a corresponding weighting function. Such integration can be done analytically, or can be treated with quadrature such that errors are reduced to rounding errors.

The Wiener chaos is an abstract representation for random variables for which convergence is clarified by the following.

Theorem 4.9 (Cameron-Martin Theorem). *The Wiener chaos representation of a random variable $X \in L^2(\Omega)$ converges in the $L^2(\Omega)$ sense to X. As a consequence of this theorem, we know the following:*

$$\lim_{p \to \infty} \int_\Omega \left| X(\omega) - \sum_{|\alpha|=0}^{p} a_\alpha \Psi^\alpha(\xi(\omega)) \right|^2 d\Pi = 0\,.$$

The Cameron-Martin theorem ensures that every random variable with finite variance has a representation in the Wiener chaos but gives no information about the convergence rate of the representation. The convergence rate is important when the series expansion is cut after a finite number of terms, which is necessary for numerical algorithms dealing with polynomial chaos expansions. In fact, [116] showed that the convergence rate of the Wiener chaos is substantially less than the optimal, exponential convergence rate. The development of other chaos types leads to expansions that have better convergence properties. This is the topic of the next section, which introduces the generalized polynomial chaos expansion.

Generalized Polynomial Chaos. Xiu and Karniadakis [116] generalized the idea of the representation of random variables in an orthogonal basis formed by polynomials in random variables with known distribution in an approach known as the gPC. They proposed to use polynomials whose weighting functions correspond to the PDF of the underlying random variables, which are the polynomials from the Askey scheme [5]. Table 4.1 shows the correspondence between important random variables and the associated polynomials.

Thus, a random variable $X \in L^2(\Omega)$ has a representation using polynomial chaos such as (4.11) and where the multidimensional polynomials Ψ^α are constructed from 1D polynomials via

$$\Psi^\alpha = \prod_{i=1}^{n} H_{\alpha_i}(\xi_i)\,, \tag{4.16}$$

Table 4.1: Important distributions and the corresponding polynomials for the expansion

Random variable	Wiener-Askey chaos	Support
Gaussian	Hermite Polynomials	$(-\infty, \infty)$
Gamma	Laguerre Polynomials	$[0, \infty)$
Beta	Jacobi Polynomials	$[a, b]$
Uniform	Legendre Polynomials	$[a, b]$
Poisson	Charlier Polynomials	discrete
Binomial	Krawtchouk Polynomials	discrete

where $H_{\alpha_i}, i = 1, \ldots, n$ are polynomials in one random variable, e.g., the ones from Table 4.2, which shows the first eight 1D polynomials for the Legendre chaos and for the Hermite chaos.

Table 4.2: The first eight 1D Legendre polynomials (*left*) and Hermite polynomials (*right*) normalized to yield orthonormal bases for the polynomial chaos

$$H_0(\zeta) = 1$$
$$H_1(\zeta) = \sqrt{3}\,\zeta$$
$$H_2(\zeta) = \frac{\sqrt{5}}{2}(3\zeta^2 - 1)$$
$$H_3(\zeta) = \frac{\sqrt{7}}{2}(5\zeta^3 - 3\zeta)$$
$$H_4(\zeta) = \frac{\sqrt{9}}{8}(35\zeta^4 - 30\zeta^2 + 3)$$
$$H_5(\zeta) = \frac{\sqrt{11}}{8}(63\zeta^5 - 70\zeta^3 + 15\zeta)$$
$$H_6(\zeta) = \frac{\sqrt{13}}{16}(231\zeta^6 - 315\zeta^4 + 105\zeta^2 - 5)$$
$$H_7(\zeta) = \frac{\sqrt{15}}{16}(429\zeta^7 - 693\zeta^5 + 315\zeta^3 - 35\zeta)$$

$$H_0(\zeta) = 1$$
$$H_1(\zeta) = \zeta$$
$$H_2(\zeta) = \frac{1}{\sqrt{2!}}(\zeta^2 - 1)$$
$$H_3(\zeta) = \frac{1}{\sqrt{3!}}(\zeta^3 - 3\zeta)$$
$$H_4(\zeta) = \frac{1}{\sqrt{4!}}(\zeta^4 - 6\zeta^2 + 3)$$
$$H_5(\zeta) = \frac{1}{\sqrt{5!}}(\zeta^5 - 10\zeta^3 + 15\zeta)$$
$$H_6(\zeta) = \frac{1}{\sqrt{6!}}(\zeta^6 - 15\zeta^4 + 45\zeta^2 - 15)$$
$$H_7(\zeta) = \frac{1}{\sqrt{7!}}(\zeta^7 - 21\zeta^5 + 105\zeta^3 - 105\zeta)$$

As a generalization of Theorem 4.9, Ernst et al. [27] proved that the generalized polynomial chaos expansion converges in quadratic mean, i.e., in the $L^2(\Omega)$ sense [47], if and only if the basic random variables have finite moments of all orders and the probability density of the basic random variables is continuous. Furthermore, the *moment problem*, i.e., the identification of the measure from the moments, has to be uniquely solvable.

To use the polynomial chaos expansion in numerical schemes, it is necessary to cut off the series expansion (4.11) after a finite number of terms. This truncation is done by choosing the number of random variables used for the approximation, denoted by n, and by prescribing the maximal polynomial degree p in the expansion, yielding an approximation of a random variable in the polynomial chaos by

$$X(\omega) \approx \sum_{|\alpha|=0}^{p} a_{\alpha} \Psi^{\alpha} (\xi(\omega)) = \sum_{|\alpha|=0}^{p} a_{\alpha} \prod_{i=1}^{n} H_{\alpha_i} (\xi_i(\omega)) =: X(\xi). \tag{4.17}$$

Again, the coefficients of this expansion are obtained through the projection (4.12). The basis functions Ψ^{α} span a space of polynomials of degree $\leq p$ in n variables that we denote by

$$\mathcal{S}_{n,p} := \mathrm{span}\{\Psi^{\alpha}\}. \tag{4.18}$$

The dimension of this space is, as usual for a polynomial basis,

$$N := \binom{n+p}{p}. \tag{4.19}$$

Note 4.10   An important theoretical result that lays foundations for using the polynomial chaos for the discretization of SPDEs is the Doob-Dynkin lemma, which states that solutions of SPDEs can be respresented in the same random variables as a finite-dimensional input that enters these equations as parameters. For details, we refer the reader to [42, 55, 110].

The remaining part of this section deals with the numerical aspects of polynomial chaos expansions. Although the presented material is valid for all polynomials from the Askey scheme, it is easier to base numerical implementations on the Legendre polynomials and uniform distributed random variables, because the support of the Legendre polynomials is compact, which is advantageous for algorithms, especially when dealing with stochastic level sets, as we will see later.

4.3.1   CALCULATIONS IN THE POLYNOMIAL CHAOS

To use the polynomial chaos in numerical schemes, it is necessary to compute moments as well as perform arithmetic operations in the polynomial chaos. Based on the work of Debusschere [20], we review the development of the basic operations such as addition, subtraction, multiplication, division, and the calculation of square roots.

In what follows, let

$$a(\xi) = \sum_{|\alpha|=0}^{p} a_{\alpha} \Psi^{\alpha}(\xi), \quad b(\xi) = \sum_{|\alpha|=0}^{p} b_{\alpha} \Psi^{\alpha}(\xi), \quad c(\xi) = \sum_{|\alpha|=0}^{p} c_{\alpha} \Psi^{\alpha}(\xi)$$

be three gPC variables of the same degree and α, β, γ multi-indices.

Expected Value, Variance, and Moments.   Because of the orthogonality of the basis functions, only the first coefficient of the expansion is relevant for the expected value, thus,

$$\mathbb{E}(a) = \int_{\Gamma} a(\xi)\, d\Pi_X = \sum_{|\alpha|=0}^{p} a_\alpha \int_{\Gamma} \Psi^\alpha(\xi)\, d\Pi_X = a_0.$$

For the higher-order moments, we also substitute the gPC expansion into the formulas from Definition 4.6 to get

$$M^n(a) = \mathbb{E}((a-a_0)^n) = \mathbb{E}\left(\left(\sum_{|\alpha|=1}^{p} a_\alpha \Psi^\alpha\right)^n\right) = \int_{\Gamma}\left(\sum_{|\alpha|=1}^{p} a_\alpha \Psi^\alpha\right)^n d\Pi_X.$$

To evaluate this integral, after the power of the sum has been expanded, the orthogonality of the basis functions is handy to simplify the computations. In particular, for the second moment (the variance), we get

$$\mathrm{Var}(a) = \int_{\Gamma}\left(\sum_{|\alpha|=1}^{p} a_\alpha \Psi^\alpha\right)^2 = \sum_{|\alpha|=1}^{p} a_\alpha^2 \int_{\Gamma} (\Psi^\alpha)^2\, d\Pi_X.$$

Sum and Difference.   We compute the sum and difference of quantities in the polynomial chaos by adding or subtracting the corresponding chaos coefficients, because the addition or subtraction of polynomials results in a polynomial with the same degree at most:

$$c = a \pm b = \sum_{|\alpha|=0}^{p} a_\alpha \Psi^\alpha(\xi) \pm \sum_{|\alpha|=0}^{p} b_\alpha \Psi^\alpha(\xi) = \sum_{|\alpha|=0}^{p} (a_\alpha \pm b_\alpha)\Psi^\alpha(\xi).$$

Product.   The multiplication of two polynomial chaos variables is more difficult. Since polynomials form the basis, the naive multiplication of the gPC variables results in a polynomial with a higher degree than the individual components. Thus, an additional projection step, here a Galerkin (L^2)-projection, onto a polynomial with the same degree as the factors of the multiplication, is necessary. The result of the projection is a gPC polynomial whose projection error is orthogonal to the space spanned by the polynomial chaos. Thus, for $c = ab$, we have for any multi-index γ

$$\int_{\Gamma} \sum_{|\alpha|=0}^{p} c_\alpha \Psi^\alpha \Psi^\gamma\, d\Pi_X = \int_{\Gamma} \sum_{|\alpha|=0}^{p} \sum_{|\beta|=0}^{p} a_\alpha b_\beta \Psi^\alpha \Psi^\beta \Psi^\gamma\, d\Pi_X$$

and so

$$c_\gamma = \sum_{|\alpha|=0}^{p} \sum_{|\beta|=0}^{p} a_\alpha b_\beta C_{\alpha\beta\gamma} \quad \text{with } C_{\alpha\beta\gamma} = \frac{\langle \Psi^\alpha \Psi^\beta \Psi^\gamma\rangle}{\langle (\Psi^\gamma)^2\rangle}. \tag{4.20}$$

Quotient.   The computation of the quotient of two random variables, $c = \frac{a}{b}$ is possible, too, for $b \neq 0$. We multiply the expression by b, yielding $a = bc$, and use again the Galerkin projection for this equation to get

$$a_\gamma = \sum_{|\alpha|=0}^{p} \sum_{|\beta|=0}^{p} C_{\alpha\beta\gamma} b_\beta c_\alpha$$

for any multi-index γ. This is a system of linear equations for the coefficients c_α, which we invert by an iterative solver.

Square Root.   In a similar manner, we compute the square root $b = \sqrt{a}$ of a polynomial chaos variable. First, we rewrite the equation in the form $a = b^2$ and then use the Galerkin projection to obtain

$$a_\gamma = \sum_{|\alpha|=0}^{p} \sum_{|\beta|=0}^{p} C_{\alpha\beta\gamma} b_\alpha b_\beta \,,$$

which is a nonlinear system of equations for the unknown coefficients b_α. We solve this nonlinear system using Newton's method to find a root of

$$f : \mathbb{R}^N \to \mathbb{R}^N, \quad f(b) = \left(\sum_{|\alpha|=0}^{p} \sum_{|\beta|=0}^{p} C_{\alpha\beta\gamma} b_\alpha b_\beta - a_\gamma \right)_{|\gamma|=0,\ldots,p}$$

in which we identify a gPC variable with the vector of its coefficients from $\mathbb{R}^N$. The partial derivatives of this function are

$$\frac{\partial f_\gamma}{\partial b_\beta}(b) = \sum_{|\alpha|=0}^{p} C_{\alpha\beta\gamma} b_\alpha \,.$$

As pointed out by Matthies and Rosic [65], it is possible to use a mild convergence criterion for Newton's method depending on the expected value and the variance of the polynomial chaos variable.

Stochastic Lookup Table.   In the above computations in the gPC, the coefficients $C_{\alpha\beta\gamma}$ are needed. Their value depends on the choice of polynomial basis and the underlying distribution of the ξ. With regard to their computation, we first note that the multidimensional integrals $\int_\Gamma \Psi^\alpha \Psi^\beta \Psi^\gamma \, d\Pi_X$ can be replaced by multiple 1D integrations. Indeed, according to (4.16), the Ψ^α are products of the polynomials in one variable. Furthermore, the random variables ξ_i are independent, thus $\mathbb{E}(\xi_i \xi_j) = \mathbb{E}(\xi_i)\mathbb{E}(\xi_j)$. In total, we get

$$\langle \Psi^\alpha \Psi^\beta \Psi^\gamma \rangle = \int_\Gamma \left(\prod_{i=1}^{n} H_{\alpha_i}(\xi_i) \right) \left(\prod_{j=1}^{n} H_{\beta_j}(\xi_j) \right) \left(\prod_{k=1}^{n} H_{\gamma_k}(\xi_k) \right) d\Pi_X$$

$$= \prod_{i=1}^{n} \int_{\Gamma_i} H_{\alpha_i}(\xi_i) H_{\beta_i}(\xi_i) H_{\gamma_i}(\xi_i) \rho_{\xi_i} \, dx \,.$$

(4.21)

In this expression, $\rho_{\xi_i}\,dx$ denotes integration with respect to the probability measures of the random variables ξ_i. It is also worth noting that

- all the integrals that appear in (4.21) can be computed exactly using an appropriate quadrature formula;

- for the fixed polynomial basis, the values can be put in a lookup table to speed up the calculations in the gPC chaos significantly; and

- for reasons of orthogonality, many of the coefficients vanish, further speeding up and simplifying computations in the gPC.

Figure 4.3 shows the sparsity structure of the tensor $C_{\alpha\beta\gamma}$. Note that the order of the tensor $C_{\alpha\beta\gamma}$ reflects the stochastic nonlinearity: Here, two gPC quantities are being multiplied. If three gPC quantities are multiplied, we end up with coefficients $\mathbb{E}(\Psi^\alpha\Psi^\beta\Psi^\gamma\Psi^\delta)$ and a 4-tensor can be used to store the precomputed values. In general, if n gPC quantities are combined by product, quotient, or square root, an $n+1$ tensor of expectations of products of stochastic basis functions results.

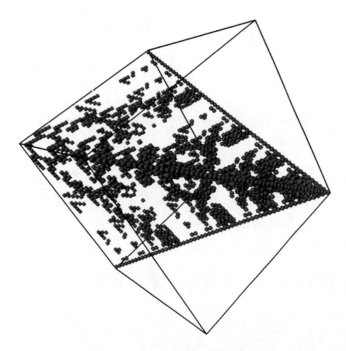

Figure 4.3: Sparsity structure of the stochastic lookup table for $C_{\alpha\beta\gamma}$ in the case of $n = 5$ random variables and a polynomial degree $p = 3$. The gray dots indicate positions in the 3D lookup table $C_{\alpha\beta\gamma}$ that contain nonzero entries.

Using the above rules, it is possible to construct numerical schemes for nearly all possible calculations. For example, the exponential of a random variable in the polynomial chaos is approximated by a truncated Taylor series expansion:

$$\exp(a) = \exp(a_1) \left(1 + \sum_{n=1}^{K} \frac{\left(\sum_{|\alpha|=1}^{p} a_\alpha \Psi^\alpha \right)^n}{n!} \right).$$

With the methods from this section, it is also possible to construct finite difference schemes for random variables as shown below.

4.3.2   POLYNOMIAL CHAOS FOR RANDOM FIELDS

We lay the foundation for a spectral discretization of stochastic PDEs by using the gPC for the stochastic part of random fields, which means that we can approximate a random field $u \in L^2(\Omega) \otimes Y(D)$ by

$$u(\xi, x) = \sum_{|\alpha|=0}^{p} u_\alpha(x) \Psi^\alpha(\xi) \tag{4.22}$$

in which the polynomial chaos coefficient u_α is now a function in $Y(D)$. Again, this mode is found by $L^2(\Omega)$ projection through

$$u_\alpha(x) = \int_\Gamma u(\xi, x) \Psi^\alpha(\xi) \, d\Pi_X . \tag{4.23}$$

Similarly, we obtain for random fields on the time-space cylinder that the modes are functions in $Y(Q)$:

$$u(\xi, t, x) = \sum_{|\alpha|=0}^{p} u_\alpha(t, x) \Psi^\alpha(\xi), \qquad u_\alpha(t, x) = \int_\Gamma u(\xi, t, x) \Psi^\alpha(\xi) \, d\Pi_X .$$

All else that was presented in this section for random variables generalizes straightforwardly to random fields on D and Q.

4.4   SAMPLING-BASED DISCRETIZATION OF SPDES

Since the middle of the twentieth century, sampling-based algorithms for the simulation of stochastic processes have been investigated, starting with the development of the Monte Carlo method. Later, more advanced sampling-based methods emerged that take into account the regularity properties of the stochastic processes. Among these approaches are the stochastic collocation method and the combination of stochastic collocation and polynomial chaos expansions or the use of sparse grids. All sampling approaches have the great advantage that they allow reusing existing deterministic code: it just has to be run with different sets of parameters, and the results need to evaluated statistically. Thus, sampling-based methods are called *nonintrusive* discretization. The disadvantage is a rather slow convergence, i.e., many samples are needed for a stable evaluation of the result's statistical properties. Much better convergence results are obtained with *intrusive* discretizations such as the stochastic FDM or stochastic Galerkin method. These methods, however, require a reimplementation of the solvers.

In what follows, we will briefly focus on sampling random fields as such fields are the objects we will be dealing with in PDE-based image processing and computer vision. This sampling is a generalization of the techniques for random variables or stochastic processes in the context of the tensor product space we introduced in Definition 4.8.

Monte Carlo Simulation.   Monte Carlo simulation is the simplest technique for the discretization of random variables and SPDEs. This technique does not take into account any assumptions on the regularity of the random variable or process such as is assumed in the polynomial chaos approximation. Thus, although the technique is simple, it is the most versatile and robust approach.

A set of samples $\xi^{(1)}, \ldots, \xi^{(R)}$ for some $R \in \mathbb{N}$ is generated randomly from the known distribution of the random variables via a pseudo random number generator such as [64]. We can use the well-known deterministic algorithms on these samples to compute depending random fields, e.g., a deterministic solver for the PDE (4.6) will yield functions $u^{(i)}(x) := u(\xi^{(i)}, x)$, $i = 1, \ldots, R$. Now, the classical formulas from statistics can be used to approximate the expected value and variance of u

$$\mathbb{E}(u)(x) \approx \bar{u}(x) = \frac{1}{R} \sum_{i=1}^{R} u^{(i)}(x) \quad \text{and} \quad \mathrm{Var}(u) \approx \frac{1}{R-1} \sum_{i=1}^{R} \left(u^{(i)}(x) - \bar{u}(x) \right)^2. \quad (4.24)$$

The main drawback of the Monte Carlo method is the slow convergence. In fact, Kendall [52] showed that the convergence of the samples' mean toward the expected value is of order $\mathcal{O}\left(1/\sqrt{R}\right)$. Despite the slow convergence rate, Monte Carlo methods are widely used (e.g., [56, 59, 89]) due to their simple implementation and the possibility to reuse deterministic code.

Stochastic Collocation.   Stochastic collocation (SC) projects the uncertainty of a random variable or process onto a space with certain properties and regularity. Again, the random field is evaluated at certain sampling points, here called *collocation points*. During the last few years, a multitude of approaches to choosing these collocation points have been developed. The approaches range from simple collocation on regular grids over sparse grid techniques to techniques allowing one to obtain polynomial chaos coefficients. We refer the reader to [114] for a more detailed overview.

In the simplest form, the collocation points are chosen following a quadrature rule, e.g., Gauss quadrature or Clenshaw-Curtis quadrature [19], where the points are selected based on the roots of the Chebyshev polynomials [107]. For multiple random variables, as in the gPC, we would need tensor product grids of Gauss or Clenshaw-Curtis collocation points with the corresponding potentiation of the number of collocation points. Using the full tensor grids in a higher dimension is known as the "curse of dimension" [73], which makes the approach practically difficult if not infeasible.

Smolyak's algorithm [82, 96] can construct a sparse grid containing significantly fewer nodes than the full tensor grid (see Figure 4.4) with an approximation order that is nearly the same as for the full tensor product grid. In fact, the approximation orders differ by only a logarithmic term (see [29]).

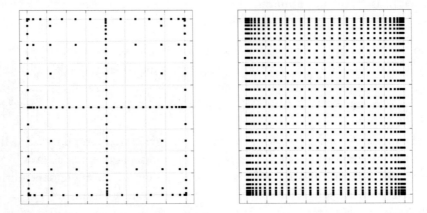

Figure 4.4: Comparison between a sparse grid (*left*) constructed via Smolyak's algorithm and a full tensor grid (*right*). The sparse grid contains significantly fewer nodes than the full tensor grid whose number of nodes grows exponentially with the dimension, but has nearly the same approximation order.

Having evaluated the random field u at $R \in \mathbb{N}$ collocation points $\xi^{(i)}$, we can compute a Lagrange interpolation polynomial in the classical way by

$$\mathcal{I}u(\xi, x) = \sum_{i=1}^{R} u(\xi^{(i)}, x) L_i(\xi) \, ,$$

where L_i are the Lagrange polynomials such that $L_i(\xi^{(j)}) = \delta_{ij}$. Using the idea of quadrature behind the selection of the stochastic collocation points, we can also evaluate integrals and thus get coefficients for a polynomial chaos expansion (cf. (4.23))

$$a_\alpha(x) = \sum_{i=1}^{R} u(\xi^{(i)}, x) \Psi^\alpha(\xi^{(i)}) w_i \, ,$$

where w_i are the corresponding quadrature weights.

4.5   STOCHASTIC FINITE DIFFERENCE METHOD

We turn now to the intrusive discretization methods. To illustrate the stochastic FDM, let us focus on a 1D version of the parabolic equation (4.10):

$$\partial_t u(\omega, t, x) - \partial_{xx} u(\omega, t, x) = f(\omega, t, x)$$

whose deterministic two-dimensional analog we have already studied in Chapter 2.

The temporal and spatial derivatives are discretized using well-known approximations analogous to the deterministic case, see Section 2.2. For example, using the explicit Euler scheme for the discretization of the time-derivative with a time-step τ, we get

$$u(\omega, t + \tau, x) = u(\omega, t, x) + \tau(\partial_{xx} u(\omega, t, x) + f(\omega, t, x)).$$

Discretizing the spatial derivative using central differences with a uniform grid width of h, the fully time-space-discrete equation reads

$$u(\omega, t + \tau, x) = u(\omega, t, x) + \tau \left(\frac{u(\omega, t, x + h) - 2u(\omega, t, x) + u(\omega, t, x - h)}{h^2} + f(\omega, t, x) \right).$$

To finish the discretization, we approximate u and f in the stochastic space by a gPC expansion, yielding a numerical scheme for the unknown modes $u_\alpha(t, x)$ at the respective nodes $(j\tau, ih) \in Q$ in the time-space cylinder Q.

Of course, any other time-stepping scheme and other finite difference stencil for the space derivatives, also in higher space dimensions, are compatible with the gPC as well. It is straightforward to derive the corresponding numerical schemes for the stochastic PDE.

4.6   STOCHASTIC FINITE ELEMENT METHOD

For a stochastic version of the well-known FEM, we closely follow the standard approach, thereby replacing quantities with their stochastic analogs as needed. As a prototype PDE, we consider the elliptic equation (4.6) and correspondingly its stochastic version (4.9).

The basis is the weak formulation of (4.6) in which we seek for $u \in V$ such that (cf. Section 2.4, (WEE))

$$A(u, v) = B(v) \quad \forall v \in V, \tag{4.25}$$

and where

$$A(u, v) = \int_D w(x) \nabla u(x) \cdot \nabla v(x) \, dx \quad \text{and} \quad B(v) = \int_D f(x) v(x) \, dx \tag{4.26}$$

are bilinear and linear forms, respectively. The space V contains all admissible functions, which also respect the Dirichlet BC, if such is prescribed as in (4.6). In such a case, $V = H_g^1(D)$ is the space of all weakly differentiable functions on D whose trace on ∂D equals the boundary function g.

For stochastic coefficients, right-hand side, or boundary conditions, the bilinear and linear forms become stochastic operators. The arguments to these operators are random fields from a tensor product space $\mathcal{S} \otimes \mathcal{V}$, e.g., the one specified in Definition 4.8 in which $\mathcal{S} = L^2(\Omega)$. We define stochastic analogs of the linear and bilinear form by also integrating over the stochastic space

$$\int_\Omega \int_D w(\omega, x) \nabla u(\omega, x) \cdot \nabla v(\omega, x) \, dx \, d\Pi = \mathbb{E}\left(\int_D w \nabla u \cdot \nabla v \, dx\right) = \mathbb{E}\left(A(u, v)\right)$$

$$\int_\Omega \int_D f(\omega, x) v(\omega, x) \, dx \, d\Pi = \mathbb{E}\left(\int_D fv \, dx\right) = \mathbb{E}\left(B(v)\right). \tag{4.27}$$

Consequently, the weak formulation of the stochastic PDE (4.9) is to find a random field $u \in \mathcal{S} \otimes \mathcal{V}$ such that

$$\mathbb{E}\left(A(u, v)\right) = \mathbb{E}\left(B(v)\right) \quad \forall v \in \mathcal{S} \otimes \mathcal{V}. \tag{4.28}$$

Thus, roughly speaking, in the weak formulation of the stochastic PDE, we take the "expected value of the deterministic weak form" by additionally integrating over the stochastic space. The solution u, and thus also the test functions v, are random fields from a tensor product space of sufficient smoothness, here $L^2(\Omega) \otimes H_g^1(D)$. To guarantee the existence and uniqueness of a solution of the weak stochastic problem (4.28), we need the form A to be continuous and coercive and the form B to be continuous on the space $\mathcal{S} \otimes \mathcal{V}$. For details, we refer the reader to [105].

From the weak form of the SPDE, we proceed to a stochastic Galerkin formulation by restricting the space $\mathcal{S} \otimes \mathcal{V}$ to a finite-dimensional version. In the stochastic component, we utilize the gPC, and thus we take $\mathcal{S}_{m,p} \subset \mathcal{S}$, and in the spatial domain, a classical finite element approach. Thus every random field $v \in \mathcal{S} \otimes \mathcal{V}$ is approximated by gPC as in (4.22) where, in addition, the modes u_α are approximated as finite element functions. Consequently,

$$u(\xi, x) = \sum_{|\alpha|=0}^{p} \sum_{i=1}^{M} u_\alpha^i P_i(x) \Psi^\alpha(\xi) \tag{4.29}$$

for scalar coefficients $u_\alpha^i \in \mathbb{R}$ and where P_i are the finite element basis functions and Ψ^α the ones from the gPC. This expression denotes a restriction of $\mathcal{S} \otimes \mathcal{V}$ to a finite-dimensional subspace $\mathcal{S}_{m,p} \otimes \mathcal{V}_h$, $\text{span}\{P_i\} =: \mathcal{V}_h$, which has dimension NM, cf. (4.19).

Restricting the weak formulation (4.28) to this space, using the Ansatz (4.29), and testing with all functions $v(\xi, x) = \Psi^\beta(\xi) P_j(x)$, we end up with

$$\mathbb{E}\left(\sum_\alpha \sum_i u_\alpha^i \Psi^\alpha(\xi) \Psi^\beta(\xi) \int_D w(\xi, x) \nabla P_i(x) \cdot \nabla P_j(x) \, dx\right) = \mathbb{E}\left(\int_D f(\xi, x) \Psi^\beta(\xi) P_j(x) \, dx\right)$$

for all multi-indices β with $|\beta| = 0, \ldots, p$ and $j = 1, \ldots, M$. We also represent the diffusion coefficient w as well as the right-hand side f in the form of (4.29) with coefficients w_γ^k and f_α^i.

If we collect the coefficients of u and f in the vectors $U_\alpha, F_\alpha \in \mathbb{R}^M$, we can write the system as

$$\sum_{|\alpha|=0}^{p} \left(L^{\alpha,\beta} \right) U_\alpha = \sum_{|\alpha|=0}^{p} M^{\alpha,\beta} F_\alpha \quad \text{for } |\beta| = 0, \ldots, p \qquad (4.30)$$

This is actually a linear system that can be expressed in block-matrix form:

$$LU = B, \qquad \text{where } L \in \mathbb{R}^{(NM)\times(NM)}, B \in \mathbb{R}^{(NM)}.$$

It has $N \times N$ blocks each of which is an $M \times M$ matrix. The blocks correspond to the coupling of the modes α, β and the matrix of each block is a variant of the well-known stiffness- or mass-matrix, i.e.,

$$
\begin{aligned}
\left(M^{\alpha,\beta} \right)_{i,j} &= \mathbb{E} \left(\Psi^\alpha \Psi^\beta \right) \int_D P_i \, P_j \, dx \\
\left(L^{\alpha,\beta} \right)_{i,j} &= \sum_k \sum_\gamma \mathbb{E} \left(\Psi^\alpha \Psi^\beta \Psi^\gamma \right) w_\gamma^k \int_D \nabla P_i \cdot \nabla P_j \, P_k \, dx.
\end{aligned}
\qquad (4.31)
$$

Note that the entries of these matrices again involve the tensor $C_{\alpha\beta\gamma}$ whose values we have stored in a stochastic lookup table. Indeed, $\mathbb{E} \left(\Psi^\alpha \Psi^\beta \right) = C_{\alpha\beta 0}$ and $\mathbb{E} \left(\Psi^\alpha \Psi^\beta \Psi^\gamma \right) = C_{\alpha\beta\gamma} C_{\alpha\beta 0}$. Furthermore, we emphasize that the formulas for the matrix blocks can, of course, be simplified if particular basis functions, such as piecewise multilinear functions on a regular grid, and quadrature rules, such as lumping of the diffusion coefficient or the masses, are chosen; see Section 2.4.

The block matrix inherits properties such as symmetry and definiteness from the properties of the deterministic matrices. Consequently, it can be solved with an iterative solver such as a (preconditioned) conjugate gradient method.

For the SFEM, existing deterministic code can be only partially reused. Depending on the regularity of the diffusion coefficient w, in the best case, the stiffness matrix blocks may be only weighted sums of the existing deterministic matrices. In the worst case, all entries need to be recomputed using the stochastic lookup table. The entries of the mass matrix blocks are simple scalings of the deterministic mass matrix. In any case, the SFEM is an intrusive discretization approach.

4.7   GENERALIZED SPECTRAL DECOMPOSITION

Figure 4.1 compares various discretization methods presented in this chapter with respect to the implementation effort and the speed of the methods. The sampling-based methods—Monte Carlo simulations and stochastic collocation—are easy to implement due to the possibility of reusing existing code. The drawback of these methods is their slow convergence toward the stochastic solution. The intrusive stochastic FEM requires significant implementation effort because it cannot reuse existing deterministic code. A significant speed-up of the solution process and an enormous reduction of memory requirements in the stochastic FEM can be achieved by selecting special subspaces and corresponding bases of $\mathcal{S}_{n,p} \otimes \mathcal{V}_h$. This choice leads to the generalized spectral decomposition (GSD) [72], which is even more elaborate and thus more difficult to implement. The advantage is the fast calculation of the stochastic result compared to the sampling-based approaches and in comparison to standard iterative solvers applied to the SFEM.

In the GSD, we approximate the solution u by

$$u(\xi, x) \approx \sum_{j=1}^{K} \lambda^j(\xi) U_j(x), \qquad (4.32)$$

where $U_j \in \mathcal{V}$ are deterministic functions, $\lambda^j \in \mathcal{S}$ are stochastic functions, and K is the number of modes of the decomposition. In this expansion, the deterministic and the stochastic basis functions are not fixed a priori. The goal of the GSD is to find an approximation to the solution with significantly fewer modes, i.e., $K \ll N$, but nearly the same approximation quality.

The representation of the approximate solution shall be the best possible in the sense of the energy norm. Thus,

$$\left\| u - \sum_{j=1}^{K} \lambda^j U_j \right\|_A^2 = \min_{\substack{\gamma^j \in \mathcal{S} \\ W_j \in \mathcal{V}}} \left\| u - \sum_{j=1}^{K} \gamma^j W_j \right\|_A^2, \qquad (4.33)$$

where the energy norm is defined as (cf. (4.26)),

$$\|v\|_A^2 := \mathbb{E}\left(A(v, v)\right) \quad \text{for } v \in \mathcal{S} \otimes \mathcal{V}.$$

The characterization of the representation is similar to what is known for deterministic problems in the context of the conjugate gradient (CG) method (or more generally in the context of Krylov space methods). Also, there is some similarity to the well-known Karhunen-Loève expansion, which has the same optimality condition in a different norm and for a known u.

The challenge given by condition (4.33) is that the solution u is not known. In the deterministic context, this challenge is circumvented by working with the residuals instead and by iteratively improving the approximation through some recursion. That is, the best approximation of order K is computed from the best approximation of order $K - 1$.

Such a recursive definition cannot be easily achieved in the stochastic case, but numerical tests show that with a recursive approach, we can achieve good approximations as well. Thus, we define the functional

$$\mathcal{J}(v) := \mathbb{E}\left(\frac{1}{2} A(v, v) - B(v)\right)$$

and we note that minimizing (4.33) is equivalent to minimizing $\mathcal{J}\left(\sum_{j=1}^{K} \gamma^j W_j\right)$. For an iterative approach, suppose that $(\lambda^1, \ldots, \lambda^{K-1})$ and $(U_1, \ldots, U_{K-1})$ have been obtained as best approximation, i.e., minimizers of $\mathcal{J}$, of order $K - 1$. Then the best approximation of the next higher-order K is obtained by

$$(\lambda_K, U_K) = \underset{\gamma \in \mathcal{S}, W \in \mathcal{V}}{\operatorname{argmin}} \mathcal{J}\left(\gamma W + \sum_{j=1}^{K-1} \lambda^j U_j\right).$$

According to Nouy [72], good results are achieved when this functional is minimized by alternately fixing the stochastic and the spatial part and by iterating this procedure until a fixed point has been reached. We introduce the residual

$$\tilde{B}(v) := B(v) - A\left(\sum_{j=1}^{K-1} \lambda^j U_j, v\right)$$

and formulate the optimality condition to find U_L for fixed λ^K as

$$\mathbb{E}\left(\lambda^K A(U_K, W)\lambda^K\right) = \mathbb{E}\left(\lambda^K \tilde{B}(W)\right) \qquad \text{for all } W \in \mathcal{V}. \tag{4.34}$$

Analogously, the optimality condition to find λ_K for fixed U_K is

$$\mathbb{E}\left(\gamma A(U_K, U_K)\lambda^K\right) = \mathbb{E}\left(\gamma \tilde{B}(U_K)\right) \qquad \text{for all } \gamma \in \mathcal{S}. \tag{4.35}$$

Thus, the minimization problem (4.33) has been transformed into a sequence of simpler minimization problems (4.34) and (4.35).

4.7.1 USING POLYNOMIAL CHAOS WITH GSD

Taking the finite-dimensional approximation with generalized polynomial chaos and finite elements into account, we arrive at a concrete algorithm for the GSD. To this end, we consider the discretized versions of A and B that we had derived in (4.31). The stochastic parts λ^j of the GSD representation are then represented as vectors $(\lambda_\alpha^j)_\alpha \in \mathbb{R}^N$, and the deterministic parts U_j are represented as vectors $(U_j)_i \in \mathbb{R}^M$. Then the residual is $\tilde{B} = (\tilde{B}^\alpha)_{|\alpha|=0,\dots,p}$, where

$$\tilde{B}^\alpha := \sum_{|\beta|=0}^{p} M^{\alpha,\beta} F_\beta - \sum_{|\beta|=0}^{p} \sum_{j=1}^{K-1} L^{\alpha,\beta} \lambda_\beta^j U_j \in \mathbb{R}^M.$$

The discrete versions of the optimality conditions (4.34) and (4.35) read

$$U_K = \left(\sum_{|\alpha|,|\beta|=0}^{p} \lambda_\alpha^K L^{\alpha,\beta} \lambda_\beta^K\right)^{-1} \left(\sum_{|\alpha|=0}^{p} \lambda_\alpha^K \tilde{B}^\alpha\right) = (\lambda^K \cdot L\lambda^K)^{-1}(\tilde{B}\lambda^K) \in \mathbb{R}^M \tag{4.36}$$

and

$$\lambda_\alpha^K = \sum_{|\beta|=0}^{p} \left(U_K \cdot L^{\alpha,\beta} U_K\right)^{-1} \left(\tilde{B}^\beta \cdot U_K\right) \in \mathbb{R} \tag{4.37}$$

Algorithm 4.1 Generalized Spectral Decomposition

1: $U \leftarrow 0, \tilde{B} \leftarrow B$
2: **for** $i = 1$ to K **do**
3:     $\lambda_\alpha^i \leftarrow 1 \quad \forall \alpha$
4:     **for** $k = 1$ to k_{max} **do**
5:         $U_i \leftarrow (\lambda^i \cdot L\lambda^i)^{-1}(\tilde{B}\lambda^i)$
6:         $U_i \leftarrow U_i / \|U_i\|_A$
7:         $\lambda_i \leftarrow L^{-1}(\tilde{B}U_i)$
8:     **end for**
9:     $U \leftarrow U + \lambda^i U_i$
10:    $\tilde{B} \leftarrow \tilde{B} - L(\lambda^i U_i)$
11: **end for**

---

or equivalently $\lambda^K = L^{-1}(\tilde{B}U_K)$, where the inversions involving L have to be interpreted correctly in the sense of the blocks of L. Thus, for computing the optimal U_K, we have to invert an $\mathbb{R}^{M \times M}$ matrix, and for computing the optimal λ^K, we have to invert an $\mathbb{R}^{N \times N}$ matrix. Through the matrices $M^{\alpha,\beta}$ and $L^{\alpha,\beta}$, the computation of λ^K and U_K again benefits from the stochastic lookup table from Section 4.3.1.

In Algorithm 4.1, we show how to finally compute the generalized spectral decomposition. This is the procedure that we have used for several applications in the following chapters. As before, the vector $U \in \mathbb{R}^{NM}$, with $U = (U^\alpha)_{|\alpha|=0,...,p}, U^\alpha \in \mathbb{R}^M$, contains the approximated solution. In line 1, the solution U is reset and the residual is initialized with the right-hand-side B. We continue with a loop until the desired order K of the approximation is reached. As stated by Nouy [72], the order $K = 8$ gives sufficiently adaquate approximations for most applications. First, in line 3, the stochastic coefficients are initialized with 1. Now these λ are kept fixed and the new iterate for the spatial part U_i is computed through (4.36). This U_i is normalized with respect to the energy norm, then kept fixed, and the new λ is obtained through (4.37) in line 7. According to Nouy, it is sufficient to iterate this alternating minimization $k_{max} = 3, 4$ times. Then, in lines 9 and 10, we update U and $\tilde{B}$ to the newly found values.

Note that in (4.36) and (4.37), we have used a sloppy notation in terms of the block matrix L. In fact, the inversions in lines 5 and 7 must be interpreted in the correct way: In line 5, equation (4.34), it is the inversion of an $\mathbb{R}^{M \times M}$ matrix that results from multiplication in the blocks of L. In line 7, equation (4.37), it is the inversion of an $R^{N \times N}$ matrix that results from multiplying the blocks with the given image vector.

CHAPTER 5

Stochastic Images

The digital image acquisition process introduces noise, which corrupts images. In fact, the repeated acquisition of the same scene does not give identical images, because the noise is typically a stochastic quantity. Consequently, the application of computer vision or image processing operators to two different images of the same scene yields different results. For quantitative information that is extracted from images, it is important to know how the uncertainty in the measurement process leads to errors in the final result of the quantification.

The key to our research in this field is the identification of images with random fields that have been introduced in Definition 4.8. We represent pixels and voxels with random variables under the mild regularity assumption that these random variables have finite variance. In principle, it is possible to assume any spatial regularity, as in classical PDE-based image processing. Since, however, we focus on discretizations with the stochastic FEM, we will assume H^1-regularity in space.

In this spirit, we can easily formulate the stochastic analogs of the definitions for classical images and image sequences that we have given in Chapter 3.

Definition 5.1 A *stochastic image* is a random field $u \in L^2(\Omega) \otimes Y(D)$, where $(\Omega, \mathcal{A}, \Pi)$ is a probability space, $D \subset \mathbb{R}^d, d \in \{2, 3\}$ is some image domain, and $Y(D)$ is a real Banach space over D.

A *stochastic image sequence* is a random field $u \in L^2(\Omega) \otimes Y(Q)$, where $(\Omega, \mathcal{A}, \Pi)$ is again a probability space, Q is the space-time cylinder $[0, T] \times D$ for some finite $T > 0$ and $D \subset \mathbb{R}^d, d \in \{2, 3\}$, and $Y(Q)$ is a real Banach space over Q.

5.1 POLYNOMIAL CHAOS FOR STOCHASTIC IMAGES

As discussed in Section 3.1, it is popular in PDE-based image processing to model an image by a representation as a finite element function, as in (3.1). Taking into account the above-formulated idea of representing pixel/voxel values by random variables, we arrive at a representation of images as random fields:

$$u(\xi, x) = \sum_{i \in \mathcal{I}} u^i(\xi) P_i(x). \tag{5.1}$$

Here we assume that every pixel/voxel-value u^i depends on a vector of random variables $\xi(\omega) = (\xi_1(\omega), \ldots, \xi_n(\omega))$ and on a random outcome $\omega \in \Omega$.

We proceed further by approximating the random variables u^i by the generalized polynomial chaos (4.17), and thus

$$u(\xi, x) = \sum_{i \in \mathcal{I}} \sum_{|\alpha|=0}^{p} u_\alpha^i \Psi^\alpha(\xi) P_i(x). \tag{5.2}$$

We call the coefficients u_α^i the *stochastic modes* of the pixel/voxel i. If we fix α and look at the set $\{u_\alpha^i\}_{i \in \mathcal{I}}$, we find that this set can be identified with a classical deterministic digital image (see Figure 5.1). Thus, the stochastic modes of stochastic images can be visualized as classical images

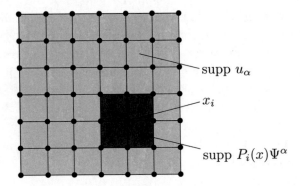

Figure 5.1: Sketch of the ingredients of a stochastic image. We discretize the spatial dimensions using finite elements, but the coefficients of the FE basis functions are random variables. Every random variable has a support, which spans over the complete image; thus pixels depend on a random vector.

as shown in Figure 5.2. These observations lead us to the following definition.

Definition 5.2   A *digital stochastic image* is a set $\{u_\alpha^i\}_{i,\alpha} \in \mathbb{R}^{|\mathcal{I}|N}$, $i \in \mathcal{I}$, $|\alpha| = 0, \ldots$, which can be interpreted as a set of deterministic digital images $\{u_\alpha^i\}_{i \in \mathcal{I}}$ that represent the stochastic modes for all α.

An analog definition holds for a *digital stochastic image sequence*.

Of course, as for deterministic digital images and image sequences, it is possible to identify a digital stochastic image and image sequence with the corresponding tensor from $\mathbb{R}^{N \times n_x \times n_y \times n_z}$ and $\mathbb{R}^{N \times n_s \times n_x \times n_y \times n_z}$ in the case of 3D images and image sequences and correspondingly for 2D images/sequences. In the context of the FDM or finite element discretization from Sections 4.5

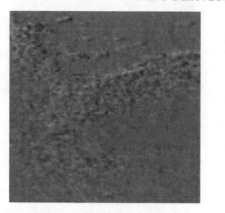 

Figure 5.2: Second (*left*) and fifth (*right*) modes of a stochastic ultrasound image. The information encoded in these images is hard to interpret because there is no deterministic equivalent.

and 4.6, we will collapse the spatial coordinate directions into one vector, e.g., by using the classical lexicographical ordering of the unknowns. Thus, $|\mathcal{I}| = M := n_x n_y n_z$ for 3D images and analogously in the case of 2D images and 2D/3D image sequences.

From the polynomial chaos expansion of a stochastic image (5.2), we compute its stochastic moments. Since $\{\Psi^\alpha, |\alpha| \le p\}$ is an orthogonal set of basis functions, we have $\mathbb{E}(\Psi^0) = 1$, $\mathbb{E}(\Psi^\alpha) = 0$ for $\alpha > 0$ and $\mathbb{E}(\Psi^\alpha \Psi^\beta) = 0$ if $\alpha \ne \beta$. Thus, we get directly from (5.2) that

$$\mathbb{E}(u)(x) = \sum_{i \in \mathcal{I}} u_0^i P_i(x) . \tag{5.3}$$

This is a deterministic image that can be identified with the set of voxel values and that can be visualized accordingly.

For the variance, the situation is a little more complicated, as the following computation shows. We have

$$\mathrm{Var}\,(u) = \mathbb{E}\big((u - \mathbb{E}(u))^2\big) = \mathbb{E}\left(\left(\sum_{i \in \mathcal{I}} \sum_{|\alpha|=1}^{p} u_\alpha^i \Psi^\alpha P_i\right)^2\right) = \sum_{i,j \in \mathcal{I}} \sum_{|\alpha|=1}^{p} u_\alpha^i u_\alpha^j \mathbb{E}((\Psi^\alpha)^2) P_i P_j .$$

Because of the product $P_i P_j$, this object is no longer an element of the space spanned by the basis functions P_k. Therefore, we need to project it back, which can be realized, for example, by simple nodal interpolation. Thus,

$$\mathrm{Var}(u)(x) \approx \sum_{i \in \mathcal{I}} \sum_{|\alpha|=1}^{p} (u_\alpha^i)^2 C_{\alpha\alpha 0} P_i(x), \tag{5.4}$$

where $C_{\alpha\alpha 0}$ is a value of the tensor from (4.20).

We obtain higher stochastic moments in a similar way, where again a projection back to the finite element space spanned by the P_i is needed. The analogous computations for image sequences are straightforward and thus not discussed further here. The objects resulting from the moment computations are deterministic images or image sequences, and thus it is possible to visualize the complete stochastic information of every pixel/voxel with the stochastic moment images or stochastic moment image sequences.

Let us emphasize that it is possible to combine the concept of stochastic images with other spatial discretizations, e.g., finite difference schemes. Then we have a pointwise representation $u(x_i, \xi)$ and corresponding equations for the stochastic moments. Numerically, we can apply the stochastic FDM as discussed in Section 4.5.

5.2 STOCHASTIC IMAGES FROM SAMPLES

The first step, required before computer vision operators can be applied on stochastic images, is the identification of the random variables ξ in the input data and the computation of the polynomial chaos coefficients u_α^i. We estimate the random variables from data samples through Karhunen-Loève expansion [22], which is a stochastic version of the well-known principal component analysis (PCA) [49]. It determines the eigenvalues and eigenvectors of the covariance matrix of given data samples and identifies the significant random variables with these eigenvectors and eigenvalues.

Let us denote by $u^{(1)}, \ldots, u^{(J)} \in \mathbb{R}^M$, $M := |\mathcal{I}|$, for some $J > 0$ digital images that have been obtained from measurements of the same object, e.g., images resulting from repeated image acquisitions. The goal is to find a vector of independent random variables $X = (X_1, \ldots, X_n)$, where $n \ll M$, that describes the distribution of the spatially dominant components of the samples.

From the input images, we can compute their mean $\bar{u}$ and obtain the $M \times M$ covariance matrix by

$$C := \frac{1}{J-1} \sum_{k=1}^{J} (u^{(k)} - \bar{u})(u^{(k)} - \bar{u})^T . \tag{5.5}$$

The eigenvalue and eigenvector pairs of C, we denote by (s_j, U_j) for $j = 1, \ldots, M$ when they are sorted in descending order. The eigenvalues U_j are images that contain the parts of the samples of varying importance/dominance.

We are searching for random variables X_j that describe the distribution of these parts of varying dominance. From the empirical Karhunen-Loève decomposition [60], we learn that

$$Y_j^{(k)} = \frac{1}{\sqrt{s_j}} U_j \cdot (u^{(k)} - \bar{u}) \in [0, 1] \tag{5.6}$$

are samples of the desired vector of random variables. However, these are samples of uncorrelated random variables that are not necessarily independent. We have to project these samples onto independent random variables to use them in the generalized polynomial chaos.

If we use Gaussian random variables as the Ansatz space for the polynomial chaos, we end up with independent random variables since uncorrelated Gaussian random variables are independent. Gaussian random variables, however, have the drawback of the infinite support of the density function, which causes problems in numerical schemes. We may also use other distributions and assume the independence as well, which leads to a small additional error, because we neglect correlation effects. However, Stefanou et al. [97] justified the assumption of independence by numerical experiments.

As a standard textbook result, we recall that if ξ has a standard uniform distribution and Y_j is a random variable with finite variance, then $F_{Y_j}^{-1}(\xi)$ has the same distribution as Y_j, i.e.,

$$Y_j = F_{Y_j}^{-1}(\xi),$$

where $F_{Y_j}^{-1}$ is the inverse of the cumulative distribution function (CDF) of the Y_j. However, in our case, we do not have the CDFs but only the empirical CDFs,

$$F_{Y_j}(y) = \frac{1}{M} \sum_{k=1}^{M} I\left(Y_j^{(k)} \leq y\right),$$

that result from the samples (5.6). Here, I is the indicator function attaining a value of 1 for true arguments and 0 else. Thus, we obtain staircase-like approximations of the random variables through

$$F_{Y_j}^{-1}(x) = \min \left\{ y \in \left\{Y_j^{(k)}\right\}_{k=1}^{M} \,\middle|\, F_{Y_j}(y) \geq x \right\}.$$

Now we may project the random variables Y_j on an element of the polynomial chaos basis by L^2 projection as shown in (4.12) and (4.23), which yields

$$X_\alpha^j = \int_\Omega F_{Y_j}^{-1}(\xi) \Psi^\alpha(\xi) \, d\Pi \qquad \text{for } |\alpha| = 0, \ldots, p. \tag{5.7}$$

Note that in this equation, the assumption of independence allows us to use basis functions, which depend on one random variable only, i.e., $\Psi^\alpha(\xi) = \Psi^\alpha(\xi_i)$. We emphasize that the assumption of independence is strong and in general not true. Still, following [97], the assumption is reasonable if the number of input samples is small. When the assumption of independence fails, it is possible to get the polynomial chaos representation via the resolution of an optimization problem on a Stiefel manifold [46], which is time-consuming. We refer the reader to Desceliers [22] and Stefanou [97] for more details about the theoretical background.

To evaluate the integral in (5.7), we can use a quadrature rule that is associated with the density $\rho(\xi)$ and thus obtain

$$X_\alpha^j \approx \sum_{k=1}^{R} w_k F_{Y_j}^{-1}\left(\xi^{(k)}\right) \Psi^\alpha(\xi^{(k)})$$

for R quadrature nodes $\xi^{(k)}$ with weights w_k. We arrive at the final representation of the image samples in terms of the polynomial chaos representation by interpolating the eigenvectors $U_j = (U_j^1, \ldots, U_j^M)$ in the finite element representation from Section 3.1, i.e.,

$$U_j(x) = \sum_{i \in \mathcal{I}} U_j^i P_i(x).$$

We then arrive at the final representation of the samples $u^{(k)}$ in terms of a random field as

$$X(\xi, x) = \sum_{i \in \mathcal{I}} \sum_{j=1}^{M} \sum_{|\alpha|=0}^{p} U_j^i X_\alpha^j \Psi^\alpha(\xi) P_i(x). \tag{5.8}$$

This equation means that the final coefficients for the representation are $\sum_j U_j^i X_\alpha^j$.

It is necessary to store only a few leading eigenvalues and eigenvectors of the covariance to capture the significant stochastic effects in the input data. Figure 5.3 shows the decay of the eigenvalues of the covariance matrix computed with 45 samples from an ultrasound device. The largest eigenvalue is associated with the mean. The other two large eigenvalues are most likely due to the motion of objects in the images during the acquisition. The higher-order stochastic effects take place on scales that are some orders of magnitude lower than the expected value.

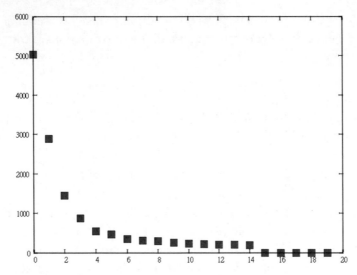

Figure 5.3: Decay of the sorted eigenvalues of the centered covariance matrix of 45 input samples from ultrasound imaging.

5.2.1 LOW-RANK APPROXIMATION

The computation and storage of the covariance matrix of the input samples is challenging because the covariance matrix is typically dense, and the memory consumption is the square of the memory consumption of a single input sample. The full storage of this matrix would limit the usability for high-resolution images enormously. To avoid the generation of the complete covariance matrix, we use the low-rank approximation developed by Harbrecht et al. [39]. This approximation is based on the pivoted Cholesky decomposition and an additional postprocessing step to generate a smaller matrix with the same leading eigenvalues.

We factorize the covariance matrix $C \in \mathbb{R}^{M \times M}$ from (5.5), into $C = LL^T$, where L is a lower triangular matrix. However, instead of the classical full factorization [34], which would require $\mathcal{O}(M^3)$ operations, we compute the factorization for a rank-m, $m \ll M$, approximation C_m. We introduce a pivot search, which guarantees that the incomplete decomposition has the same leading eigenvalues as the original matrix. Then a rank m approximation is given by the product of the two Cholesky factors L_m and L_m^T, i.e., $C_m = L_m L_m^T$, where the Cholesky factors are computed using Algorithm 1 from [39]. This algorithm needs access to the diagonal of the matrix C and m rows of the matrix only. The storage requirement thereby decreases from M^2 to $(m + 1)M$ and the number of operations from $\mathcal{O}(M^3)$ to $\mathcal{O}(m^3)$. However, this algorithm computes the exact values for the leading eigenvalues, not approximations.

The computation of the eigenvalues of C_m benefits from the fact that the eigenvalues of $C_m = L_m L_m^T$ are the same as the eigenvalues of $\tilde{C}_m = L_m^T L_m$. Thus, we transform the computation of the m leading eigenvalues from a $\mathbb{R}^{M \times M}$ matrix into the computation of the m eigenvalues of a $\mathbb{R}^{m \times m}$ matrix, where $m \ll M$. The eigenvectors of the initial matrix C are $x = L_m \hat{x}$, where $\hat{x}$ are the eigenvalues of the small matrix $L_m^T L_m$. We refer to Harbrecht [39] for details about the theoretical background.

5.3   STOCHASTIC IMAGES FROM NOISE MODELS

The generation of stochastic input data from image samples is not feasible for cases in which multiple samples simply cannot be recorded because the desired imaging modality does not allow for such (cf. Section 5.4). In this situation, the single image that has been acquired needs to serve as an estimate for the first moment, i.e., the expected value.

With a noise model (see e.g., [93]), we can equip every pixel/voxel of the estimated expected value with a distribution. If the noise model has a distribution that corresponds exactly to one of the possible polynomial chaos bases (cf. Section 4.3 and Table 4.1), the coefficients of the gPC representation are conveniently found. For other noise distributions, a Galerkin projection, as described in Section 4.3, can be used to find the gPC representation of the stochastic input image.

This procedure, however, leads to high-dimensional stochastic spaces since we cannot detect correlations as in the Karhunen-Loeve approach. In fact, we have to equip every pixel/voxel with the full probability space and the full gPC expansion, thus arriving at very high-dimensional spaces. Thus, if no further assumptions on correlations or the number of random variables can be made in the actual application, the use of noise models is practicable for small images only.

5.4 IMAGING MODALITIES FOR STOCHASTIC IMAGES

What kind of imaging is available, and which method for providing stochastic images as measurements for computer vision/image processing can be used, clearly depends on the type of object and the information that shall be gained about the object. In this section, we will briefly comment on the various imaging modalities and how they may be used for generating stochastic images.

Optical Imaging. The use of optical imaging is probably the least problematic for all materials. However, it is mainly suitable for surface imaging only. For most materials, there is a low penetration depth only. The acquisition of multiple samples is not problematic.

Ultrasound Imaging. As in optical imaging, the use of ultrasound imaging is feasible for most materials. Thus, ultrasound imaging would allow us to obtain multiple samples.

Computed Tomography or X-Ray Imaging. Repeated CT imaging or X-ray imaging may be suitable for technical materials but certainly not for biological materials. The high-energy radiation used in CT and X-rays either destroys directly or poses a risk for harm that renders its intensive use unethical. We may use a noise model here. However, we can also generate stochastic images from the so-called *sinogram*, which is the raw data of the acquisition process (see, e.g., [8] for details) that requires a reconstruction step to yield the medical image. The sinogram is the collection of rays through the object under different angles and directions in CT imaging [40]. Thus, only one recording of the sinogram is necessary to obtain multiple samples with different reconstruction algorithms as the input for stochastic computer vision and image processing. Our approach is based on the hypothesis that the sinogram (see Figure 5.4) is free of noise and that the noise and the artifacts in the final CT images are due to the reconstruction step, which is necessary to transform the sinogram into the final data set. Thus, we may use multiple reconstruction techniques and parameter settings to generate the input samples and the technique described in the previous section to generate the stochastic images. The reconstruction techniques range from Fourier-based methods to iterative methods with different settings for the data interpolation and the filter window for the low-pass filtering [109]. For the different reconstructions, we use CTSim [90], for which source code is available. In our work, we have combined the generation of input samples and the computation of the resulting stochastic image in one program that runs without user interaction.

Magnetic Resonance Imaging. For biological samples it can be difficult to obtain multiple images from MR imaging. Thus, this imaging modality is a candidate for noise models as described above. For example, the distribution of noise in MR imaging is known to be Rician [38]. Another option is to use different reconstruction algorithms as well. MRI data are acquired in the so-called *k-space* [104] and require a reconstruction to the physical space as for the CT sinogram.

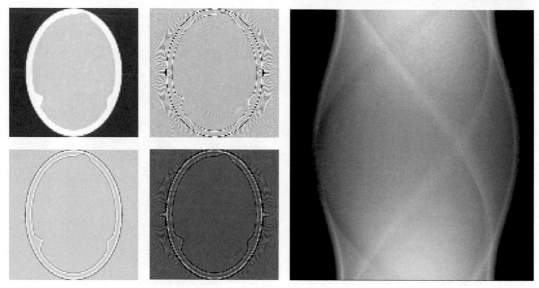

Figure 5.4: *Left picture group, from left to right and top to bottom:* The first mode (=expected value), second mode, third mode, and fourth mode of a stochastic CT image. (*right*) The sinogram, i.e., the raw data produced by the CT imaging device for the head phantom [95].

5.5 VISUALIZATION OF STOCHASTIC IMAGES

The visualization of uncertainty has been an active research topic for several years, see, e.g., [37, 84] and the references therein. As the resulting tensor product spaces of spatial and stochastic components are of very high dimensions, the intuitive visualization faces enormous challenges. With today's techniques, either the spatial or the stochastic dimensions need to be reduced, for example, by showing few stochastic modes or moments.

Stochastic Images. For 1D data, it is possible to draw additional information in the graph of the function, e.g., displaying the standard deviation and other stochastic quantities such as kurtosis or skewness [84]. The stochastic images introduced here are 2D or 3D with possible additional temporal dimensions for image sequences. We have to visualize these space dimensions and the additional stochastic dimensions in terms of the polynomial chaos expansion.

A stochastic image is given by (5.2) and, thus, the visualization techniques for classical images are only partially feasible. One possibility for the visualization is to separately display the stochastic modes, i.e., the coefficients of the polynomial chaos expansion (5.2), via the images shown in Figure 5.2. There the set $f_\alpha^i, i \in \mathcal{I}$ for fixed α is visualized as a single classical image. The complete stochastic image can be shown as N of such images, which is disappointing for images with high-stochastic dimensions.

A second option is to show stochastic moments as classical images. Thus we would compute the stochastic moments from the modes through (5.3), (5.4), and analog formulas. In Figure 5.5, the expected value and variance are shown, which allows us to get an impression of the variability of the pixels.

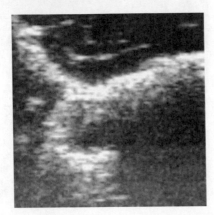 

Figure 5.5: Expected value (*left*) and variance (*right*) of a stochastic ultrasound image. The expected value looks like a deterministic image. In the variance image, regions of the stochastic image that have high gray value uncertainty are visible as white dots.

Thirdly, we may simply draw a set of samples from the computed output distribution. With such sampling, we display classical pictures, as shown in Figure 5.6; however, the visual impression is clearly influenced by the randomly chosen samples. For a moderate number of random variables, it is also possible to generate specific samples from stochastic images by prescribing the value for every random variable ξ_i. From these, we obtain the values of the $\psi^{\alpha}(\xi)$, with which the mode images u_{α} need to be weighted and summed up. It is also possible to generate an animation that shows possible realizations of the stochastic image by looping over the process just described [37].

Stochastic Level Sets.   For segmentation on stochastic images, it is necessary to visualize stochastic contours, i.e., contours whose positions and shapes depend on random variables. Again, in a Monte Carlo fashion, we may draw samples from the stochastic level set function and thus arrive at different realizations of the stochastic contour, i.e., the zero level set. We can visualize these realizations of the stochastic contour as shown in Figure 5.7. Then, the uncertainty is coded visually in the distance between the realizations of the sample-contours, i.e., a high uncertainty is shown by the greater distance between realizations of the contours.

Through the sampling of the stochastic contour, we may also derive some of its stochastic moments, which can be visualized as a color overlay on the expected contour.

In the segmentation of stochastic 3D images, we arrive at level sets that are stochastic surfaces. Their visualization is even harder. A plane-by-plane slicing would allow visualization

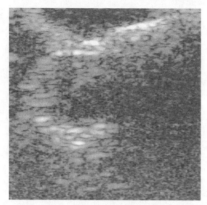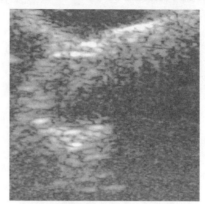

Figure 5.6: Two samples drawn from a stochastic image. The images differ due to realizations of the noise. In a printed version, these images look nearly the same.

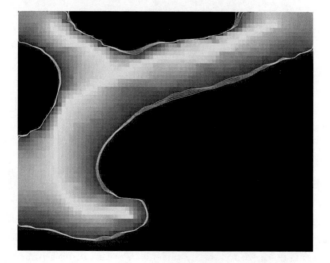

Figure 5.7: Visualization of realizations of a stochastic 2D contour. The yellow lines correspond to a realization of the stochastic contour that is encoded in the stochastic image.

of stochastic contours as described in the previous paragraph. However, this process is cumbersome and does not yield an intuitive impression of the stochastic surface. Another possibility is to visualize the expected value surface and to color-code it by the variance or higher-order moments [84]. Figure 5.8 shows such a visualization. The result is an image that is comparable to the 2D result from Figure 5.5 but that combines the information into one image. Djurcilov [23] presented further ideas for the volume rendering of stochastic images.

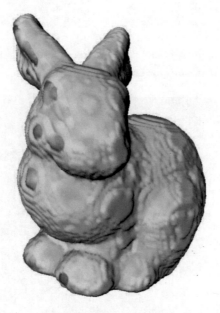

Figure 5.8: Visualization of a 3D contour encoded in a 3D stochastic image. The expected value of the 3D stochastic contour is color-coded by the variance. A color ramp from red over yellow to green indicates the range of the variance from high to low.

CHAPTER 6

Image Processing and Computer Vision with Stochastic Images

The task of this chapter is to combine the notion of stochastic images with the concept of SPDEs introduced in Chapter 4. SPDEs arise from variational formulations of image processing problems when we apply these variational methods to stochastic images. In this chapter, we investigate linear and nonlinear stochastic denoising methods; segmentation methods based on elliptic, parabolic, and hyperbolic SPDEs; and stochastic registration methods.

Based on elliptic SPDEs, we develop two segmentation methods for stochastic images: random walker segmentation and Ambrosio-Tortorelli segmentation of stochastic images. The segmentation methods differ in reference to user interaction and the number of parameters. The extension of the random walker segmentation is interactive. Thus, it is possible to improve the segmentation quality by adding additional seed regions interactively. On the other hand, the extension of the Ambrosio-Tortorelli segmentation is fully automatic. The user tunes the parameters only and has no chance of improving the quality of the segmentation afterward, except by choosing a new set of parameters.

We start the expositions with the simple linear and nonlinear diffusion filtering models, as the resulting SPDEs are very similar to the prototypic SPDEs that have the discretization that we discussed in detail in Chapter 4. To verify our implementations and to show the benefit of the general spectral decomposition, we show comparisons with the well-known and very robust Monte Carlo Method.

6.1 STOCHASTIC DIFFUSION FILTERING

From the classical diffusion filtering equations that have been described in Section 3.2, we arrive at the stochastic variants by replacing functions with random fields. Thus, the equation for linear diffusion filtering of a stochastic input image $u_0 \in L^2(\Omega) \otimes L^2(D)$ with a stochastic diffusion coefficient $w \in L^2(\Omega) \otimes L^2(D)$ is given by

$$\begin{aligned}
\partial_t u(\omega, t, x) - \mathrm{div}(w(\omega, x)\nabla u(\omega, t, x)) &= 0 \qquad \text{almost surely on } \mathbb{R}_0^+ \times D, \\
u(\omega, 0, x) &= u_0(\omega, x) \quad \text{almost surely in } D
\end{aligned} \qquad (6.1)$$

with Neumann BC. This is the stochastic heat equation that we have already discussed as a prototype of a parabolic SPDE in Section 4.2, see (4.10).

In case of a nonlinear diffusion filtering in the spirit of Perona-Malik, the stochastic generalization is given by

$$\partial_t u(\omega, t, x) - \operatorname{div}\left(g(|\nabla u_\sigma|(\omega, t, x))\nabla u(\omega, t, x)\right) = 0 \qquad \text{almost surely on } \mathbb{R}_0^+ \times D,$$
$$u(\omega, 0, x) = u_0(\omega, x) \quad \text{almost surely on } D,$$
$$(6.2)$$

where we restrict our expositions here to a Perona-Malik model with a vanishing right-hand side $f \equiv 0$. The generalization to a nonvanishing right-hand side is straightforward and can be easily realized with the concepts described in the previous chapters.

The linear diffusion equations (6.1) and (6.2) are parabolic SPDEs, which are supposed to be interpreted in the weak sense. Here, we will give some details for the Perona-Malik equation only as the discretization of the heat equation can be easily synthesized from that. Following the expositions of Section 4.2, we multiply the equation with a random field test function v and integrate the resulting product over both physical and stochastic space. This procedure leads us to

$$\int_\Omega \int_D \partial_t u(\omega, t, x)v(\omega, x) - \operatorname{div}\left(g(|\nabla u_\sigma|(\omega, t, x))\nabla u(\omega, t, x)\right) v(\omega, x)\, dx\, d\Pi = 0 \qquad (6.3)$$

for all $v \in L^2(\Omega) \otimes L^2(D)$. Replacing the time-derivative by a semi-implicit, finite difference time-stepping scheme with time-step τ and integrating by parts the second term of the equation yields

$$\int_\Omega \int_D u(\omega, t, x)v(\omega, x) + \left(g(|\nabla u_\sigma|(\omega, t-\tau, x))\nabla u(\omega, t, x)\right) \cdot \nabla v(\omega, x)\, dx\, d\Pi =$$
$$\int_\Omega \int_D u(\omega, t-\tau, x)v(\omega, x)\, dx\, d\Pi. \quad (6.4)$$

Here, we assume homogeneous Neumann BC for u such that no boundary terms appear in the weak form. As usual for the nonlinear diffusion filtering, the semi-implicit character of the time-stepping scheme results in an evaluation of the diffusion coefficient at the old time-step.

For the existence of solutions of this weak SPDE, the coefficient $g(|\nabla u_\sigma|(\omega, t, x))$ has to be positive and bounded almost surely. In fact, cf. (4.7), there must exist $w_{\min}, w_{\max} \in (0, \infty)$ such that

$$\Pi\left(\omega \in \Omega \mid g(|\nabla u_\sigma|(\omega, t, x)) \in [w_{\min}, w_{\max}] \quad \forall (t, x) \in \mathbb{R}_0^+ \times \overline{D}\right) = 1. \qquad (6.5)$$

This property is easily asserted for the usual diffusion tensors of the form $g(s) = (1 + s^2/\lambda^2)^{-1}$ for some $\lambda > 0$.

The weak system (6.4) is discretized in physical space and stochastic space by a substitution of the polynomial chaos expansion (5.2) of the images u and u_0 as well as the coefficient g. As in

Section 4.2, test functions will be products $\Psi^\beta(\xi)P_j(x)$ of spatial basis functions and stochastic basis functions. Denoting the vectors of coefficients for the stochastic modes of the new time-step with $U_\alpha = (u_\alpha^i)_{i\in\mathcal{I}} \in \mathbb{R}^M$, for the old time-step with $\bar{U}_\alpha = (u_\alpha^i)_{i\in\mathcal{I}} \in \mathbb{R}^M$, and similarly for u_0 and g, we get the fully discrete system

$$\sum_{|\alpha|=0}^{p} \left(M^{\alpha,\beta} + \tau L^{\alpha,\beta} \right) U_\alpha = \sum_{|\alpha|=0}^{p} M^{\alpha,\beta} \bar{U}_\alpha \tag{6.6}$$

for all $|\beta| = 0, \ldots, p$. This system has to be solved for every time-step $m\tau$, $m \in \mathbb{N}$, i.e., U_α represents $u(\xi, m\tau, x)$ and $\bar{U}_\alpha$ represents $u(\xi, (m-1)\tau, x)$. The matrices $M^{\alpha,\beta}$, $L^{\alpha,\beta}$, and $S^{\alpha,\beta}$ are blocks of the system matrix (see Figure 6.1). The single entries of these matrices are defined as

$$\left(M^{\alpha,\beta} \right)_{i,j} = \mathbb{E}\left(\Psi^\alpha \Psi^\beta \right) \int_D P_i P_j \, dx,$$
$$\left(L^{\alpha,\beta} \right)_{i,j} = \sum_k \sum_\gamma \mathbb{E}\left(\Psi^\alpha \Psi^\beta \Psi^\gamma \right) g_\gamma^k \int_D \nabla P_i \cdot \nabla P_j P_k \, dx. \tag{6.7}$$

The polynomial chaos representation of the stochastic edge indicator $g(\xi, x) = \sum_{k\in\mathcal{I}} \sum_{|\gamma|=0}^{p} g_\gamma^k \Psi^\gamma(\xi) P_k(x)$ is obtained by projecting the edge indicator to the polynomial chaos with the calculation schemes from Section 4.3.1. The smoothing of the stochastic image as input for the edge indicator is obtained by solving the stochastic heat equation until stopping time $T = 0.5\sigma^2$. The stiffness matrix for this linear diffusion with diffusion coefficient $w \equiv 1$ is given by

$$\left(S^{\alpha,\beta} \right)_{i,j} = \mathbb{E}\left(\Psi^\alpha \Psi^\beta \right) \int_D \nabla P_i \cdot \nabla P_j \, dx. \tag{6.8}$$

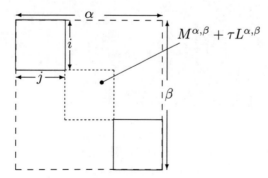

Figure 6.1: Structure of the block system of a diffusion SPDE. Every block has the sparsity structure of a classical finite element matrix, and the block structure of the matrix is sparse, meaning that some of the blocks are zero. The sparsity structure on the block level depends on the number of random variables and the polynomial chaos degree used in the discretization.

Clearly, the entries of the above matrices are weighted sums of the entries of the classical mass and stiffness matrices. The weights are given by the $C_{\alpha\beta\gamma}$, and thus the sparsity structure of this tensor induces sparsity in the block system.

6.1.1   RESULTS

To demonstrate the performance of the stochastic heat equation and the Perona-Malik selective smoothing, we study the diffusion of a square in Figure 6.2. The square has a constant mean and a gradient in its variance. We see from the figure that linear smoothing behaves as expected: both the mean and the variance are smoothed and edges are lost. With increasing time, the expected value and the variance will become constant. In contrast to traditional diffusion, the Perona-Malik diffusion retains edges in the expected value, and the variance is increased in the vicinity of the edges. The variance is smoothed also in the interior of the square.

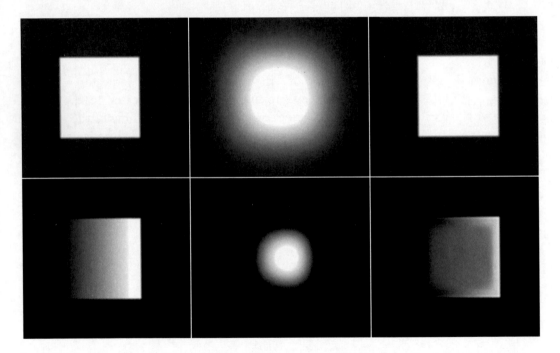

Figure 6.2: Result from stochastic linear and nonlinear diffusion of a square. The *top row* shows the expectation, and the *bottom row* shows the variance. The input image is *on the left*, the result from linear diffusion with the heat equation is *in the middle*, and the result from stochastic Perona-Malik smoothing is *on the right*. It is clearly visible how the detected edges influence the variance of the result in the Perona-Malik smoothing.

6.2   STOCHASTIC RANDOM WALKER AND STOCHASTIC DIFFUSION IMAGE SEGMENTATION

The extension of the random walker segmentation [36] (which was summarized in Section 3.3) to a stochastic segmentation method is straightforward. We replace the classical image $u : D \to \mathbb{R}$ with a stochastic image $u : \Omega \times D \to \mathbb{R}$ as defined in (5.2). Random walker segmentation needs no assumptions about the regularity of the input images because it transforms the problem into a partition problem of a graph. However, to prove the existence and uniqueness of solutions, we relate to the continuous formulation and the deduced SPDE. Further, we restrict the method to images with an H^1-regularity in the spatial dimensions.

Let us start by building a graph for the spatial dimensions of the stochastic image. On this graph, we define stochastic analogs of the edge weights and node degrees. The stochastic edge weight, a random variable, is given by the same expression as the classical edge weight (3.8), but the quantities extracted from the image are random variables. The random variable describing the weight of the edge between neighboring pixels i and j is

$$w^{ij}(\xi) = \exp\left(-\beta \left(g^i(\xi) - g^j(\xi)\right)^2\right),$$

where $g^i - g^j$ represents a normalized difference between neighboring pixels. However, for random fields g^i and g^j, the normalization step (to ensure that the maximal difference between g^i and g^j is one), is not straightforward. In terms of random variables, a normalization ensures that their expected value equals 1. Thus, we achieve the normalization by dividing the difference of neighboring pixels by the maximal difference of the expected value of neighboring pixels, i.e.,

$$\left(g^i(\xi) - g^j(\xi)\right)^2 = \frac{(u^i(\xi) - u^j(\xi))^2}{\max_{k,l} \mathbb{E}\left(\left(u^k(\xi) - u^l(\xi)\right)^2\right)}.$$

Furthermore, using the polynomial chaos representation of u, we obtain

$$w^{ij}(\xi) = \exp\left(-\beta \frac{\left(\sum_{|\alpha|=0}^{p} u_\alpha^i \Psi^\alpha(\xi) - \sum_{|\alpha|=0}^{p} u_\alpha^j \Psi^\alpha(\xi)\right)^2}{\max_{k,l} \mathbb{E}\left(\left(u^k(\xi) - u^l(\xi)\right)^2\right)}\right).$$

In Section 4.3, it was described how to perform these calculations for random variables represented in the polynomial chaos. Note that here we do not calculate the exponential of the polynomial chaos expansion explicitly. Instead, we compute a Galerkin projection of the exponential in the polynomial chaos to obtain the modes w_α^{ij}, $|\alpha| = 0, \ldots, p$ of the stochastic edge weights.

Now it is easy to build the stochastic analog of the Laplacian matrix (3.7) as

$$
M_{ij}(\xi) = \begin{cases} \sum\limits_{\substack{\text{neighboring} \\ \text{nodes } x_k}} w^{ik}(\xi) & \text{if } i = j, \\[1em] -w^{ij}(\xi) & \text{if } x_i \text{ and } x_j \text{ are neighboring nodes,} \\[1em] 0 & \text{else.} \end{cases}
$$

We end up with $M(\xi) = \sum_{|\alpha|=0}^{P} M_\alpha \Psi^\alpha(\xi)$ for appropriately defined moment matrices M_α. In fact, the coefficient M_α in this polynomial chaos expansion is a matrix containing at position i, j the αth coefficient of the polynomial chaos expansion of the entries $M_{ij}(\xi)$.

To define the linear system of equations to solve the stochastic random walker problem, let us now approach the model from the viewpoint of the stochastic analog of the weighted Dirichlet integral (3.5). We take the expected value of the classical weighted Dirichlet integral E_{RW} and insert the stochastic quantities, thus

$$
\mathbb{E}\left(E_{RW}\left[u(\xi, x)\right]\right) = \mathbb{E}\left(\frac{1}{2}\int_D w(\xi, x)|\nabla u(\xi, x)|^2\, dx\right). \tag{6.9}
$$

As for the classical energy (cf. Section 3.3), a minimizer is a harmonic function satisfying

$$
\begin{aligned}
-\text{div}\left(w(\xi, x)\nabla u(\xi, x)\right) &= 0 \quad \text{almost surely in } D, \\
u &= 1 \quad \text{almost surely on } V_O, \\
u &= 0 \quad \text{almost surely on } V_B,
\end{aligned} \tag{6.10}
$$

where V_O and V_B play the role of user-prescribed initialization of object and background as in the deterministic case. Remember that we define the seeded domain as $V_S = V_O \cup V_B$ and $V_U = D \setminus V_S$.

Now we move to a discrete representation of u and w by using polynomial chaos as described in the above paragraphs. Thereby, we split the representation into the boundary conditions $u_S(\xi, x) = u(\xi, x)|_{V_S}$ that result from the seeded domain and into the actual degrees of freedom $u_U(\xi, x) = u(\xi, x)|_{V_U}$. The corresponding vectors of nodal and modal values are denoted by U_U and U_S.

The discrete version of the stochastic Dirichlet integral (6.10) is expressed with the matrix M as

$$
\begin{aligned}
\mathbb{E}(E_{RW}[u]) &= \mathbb{E}\left(\frac{1}{2}U \cdot MU\right), \\
&= \mathbb{E}\left(\frac{1}{2}\begin{bmatrix} U_S \\ U_U \end{bmatrix} \cdot \begin{bmatrix} L_S & B \\ B^T & L_U \end{bmatrix}\begin{bmatrix} U_S \\ U_U \end{bmatrix}\right), \\
&= \mathbb{E}\left(\frac{1}{2}\left(U_S \cdot L_S U_S + 2U_S \cdot B U_U + U_U \cdot L_U U_U\right)\right),
\end{aligned}
$$

where we split the stochastic matrix M into $L_U = M|_{(i,j):x_i,x_j \in V_U}$, $L_S = M|_{(i,j):x_i,x_j \in V_S}$ and $B = M|_{(i,j):x_i \in V_U, x_j \in V_S}$. A stochastic minimizer of the discretized stochastic Dirichlet problem is given by

$$L_U(\xi)U_U(\xi) = -B^T(\xi)U_S(\xi). \tag{6.11}$$

This system of linear equations is solved using the generalized spectral decomposition from Section 4.7.

Due to the construction of the solution via a problem on a graph, we end up with a much simpler stochastic problem in comparison to a direct solution of the SPDE (6.10). Using the definition of the solution via the graph, the matrix M has a block diagonal representation $M = \text{diag}(M^1, \ldots, M^N)$. If we discretize the SPDE (6.10) with an SFEM approach, we end up with a matrix that has nonzero blocks away from the diagonal. This difference is due to a projection step of the graph representation, because the random variables $w^{ij}(\xi)$ are projected back to the polynomial chaos at an early stage. In contrast, when using the SFEM, this projection would take place at the end of the solution process.

6.2.1   RESULTS

We demonstrate the benefits of the stochastic extension on three data sets. The first data set consists of $M = 5$ samples with a resolution of 100×100 pixels from the artificial "street sequence" [66]. Note that we do not consider the images as a sequence; instead, we treat them as samples of the noisy and uncertain acquisition of the same scene. The second data set consists of 45 ultrasound (US) samples with a resolution of 300×300 pixels. The third data set is a liver mask with varying background and a resolution of 129×129 pixels. The image is corrupted by uniform noise, and 25 samples with different noise realizations are treated as input. We compute a stochastic image containing $n = 5$ random variables for the ultrasound device, $n = 3$ for the liver, and $n = 2$ for the street scene. The number of random variables is chosen based on the exponential decay of the eigenvalues of the covariance matrix of the input samples (cf. Section 5.2). It is sufficient to store a few of the random variables to capture the important stochastic effects. For the three data sets, we use a polynomial degree $p = 3$ and we compute a stochastic image from these samples using the method from Section 5.2. This polynomial degree is a good balance between accuracy and computational effort.

The user defines the seed points for segmentation on the expected value of the stochastic image (see Figure 6.3). With the stochastic image and the seed points as input, we perform the stochastic random walker segmentation. The parameter β varies during the experiments. Together with the mean, we are able to show the segmentation variance per pixel. The variance of the polynomial chaos expansion of the result contains information about how the gray value uncertainty in the input image propagates through the segmentation operator and influences the result. Regions with a high variance indicate where the input gray value uncertainty influences the object detection. The uncertainty changes from the input to the output. In the input data, the gray value uncertainty spreads over the whole image, whereas in the segmentation result, the gray value uncertainty concentrates at the object boundary.

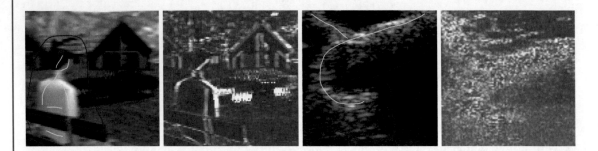

Figure 6.3: Expected value (*left*) and variance (*middle left*) of the street image; expected value (*middle right*) and variance (*right*) of the US image. The seed regions are color-coded for the interior (*yellow*) and exterior (*red*).

Note that the classical random walker result can be interpreted as a probability map, i.e., the result of random walker segmentation is a probability for every pixel to belong to the object or not. When we apply the stochastic random walker method, the first interpretation of the result is that we compute "probabilities of probabilities" because we are computing the probability distribution of the values at every pixel. We emphasize that this is only one possible interpretation of the results of the classical random walker method.

In fact, we are computing the result of a diffusion problem, and the stochastic extension is a diffusion problem with a stochastic diffusion coefficient. The results can be interpreted as the stochastic solution of the stochastic diffusion problem. The analog to the probability interpretation is that we compute the probabilities for belonging to the object in dependence on the input gray value uncertainty.

The stochastic boundary can be visualized by tracking the deterministic object boundary (i.e., the value 0.5 in the result image) for realizations of the random variables. In this visualization, we are inspired by the work of Prassni et al. [85]. The difference is that Prassni et al. [85] visualize the isolines of different probabilities whereas we visualize the same isoline for realizations of the stochastic image. Figure 6.4 shows the result of such a visualization.

It is possible to compute and visualize other quantities extracted from the segmentation, e.g., the volume of the segmented object. The straightforward visualization of the stochastic volume is to draw the PDF of the volume. The PDF can be computed from the segmentation by summing up the random variables at every pixel, because they specify the "probability" that the pixel belongs to the object. Thus, given the random walker solution u, which can be interpreted as an indicator function for the object, the random variable $v(\xi)$ specifying the object's volume

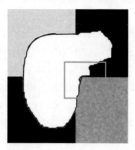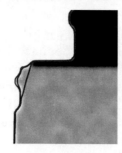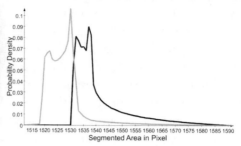

Figure 6.4: (*left*) Monte Carlo realizations of the stochastic object boundary with $\beta = 10$. (*center*) we highlight an image region where the noise in the input image influences the result. (*right*) PDF of the segmented person area from the street image for $\beta = 25$ (black) and $\beta = 50$ (gray). From the PDF, we judge the segmentation, reliability. A narrow PDF indicates that the image noise influences the segmentation marginally.

is

$$v(\xi) = \sum_{|\alpha|=0}^{p} v_\alpha \Psi^\alpha(\xi) := \sum_{i \in \mathcal{I}} u^i(\xi) = \sum_{|\alpha|=0}^{p} \sum_{i \in \mathcal{I}} u_\alpha^i \Psi^\alpha(\xi). \tag{6.12}$$

Looking at the PDF, it is easy to decide whether the image noise influences the segmented volume strongly or not. If the segmented volume is strongly influenced, the PDF is broad; otherwise, the function is narrow. Figure 6.4 shows the PDF of the segmented volume from the street image. Also, the choice of the parameter β influences the profile of the PDF. A smaller β leads to a diffuse object boundary and a broader PDF. The dependence and sensitivity of computer vision models on their parameters, of which we see an example here, will be discussed further in Chapter 7.

Another possibility to calculate the object volume is to count pixels with a value above 0.5 only. To this end, we compute image samples from the stochastic result via a Monte Carlo approach, threshold these samples, count the number of object pixels, and calculate the PDF of the volume. This method, however, has two drawbacks: it is time-consuming and it does not consider partial volume effects [8] at the boundary because it considers pixels with a probability of more than 0.5 only.

6.2.2   PERFORMANCE EVALUATION

The stochastic random walker model allows us to impressively demonstrate the varying performances of the different discretization approaches. We show a detailed comparison only for the random walker segmentation, but the results generalize to the Ambrosio-Tortorelli level set segmentation and other models that use the same discretization methods.

Table 6.1 shows the comparison of the execution times of the GSD method, the Monte Carlo method, and the stochastic collocation method with a sparse Smolyak grid and a full tensor product grid. It is easy to see that the GSD method outperforms all sample-based approaches. Thus, we prefer the GSD method and the FDM in all our SPDE computer vision models. The stochastic collocation methods suffer from the "curse of dimension" [73] because the execution times grow exponentially with the number of random variables in the stochastic images.

On the other hand, sampling methods are much easier to implement and existing code can be reused. One further benefit of the Monte Carlo method is its independence from the number of random variables. Nevertheless, the 1,000 samples used here are a lower bound for the number of runs needed to get accurate results. Recall that (cf. Section 4.4) the rate of convergence is $\mathcal{O}\left(1/\sqrt{R}\right)$, where R is the number of samples, and even with this "small" number of runs, the Monte Carlo is slower than GSD.

Table 6.1: Comparison of the execution times (in seconds) of the discretization methods

	Street ($n = 2$)	Liver ($n = 3$)	Ultrasound ($n = 5$)
Monte Carlo (1,000 samples)	76	113	1,814
Stochastic Collocation (full grid)	16	390	$\approx 1,400,000$
Stochastic Collocation (sparse grid)	6	18	634
GSD	9	15	437

6.3   STOCHASTIC AMBROSIO-TORTORELLI SEGMENTATION

For the segmentation of stochastic images by the phase field approach of Ambrosio and Tortorelli [4], we replace the deterministic u and ϕ with their stochastic analogs. The stochastic energy components are defined as expectations of the classical energies (cf. Section 3.4), i.e.,

$$E_{\text{fid}}^s[u] := \mathbb{E}(E_{\text{fid}}[u]) = \int_\Omega \int_D (u(\omega, x) - u_0(\omega, x))^2 \ dx \, d\Pi$$
$$E_{\text{reg}}^s[u, \phi] := \mathbb{E}(E_{\text{reg}}[u, \phi]) = \int_\Omega \int_D \mu \left(\phi(\omega, x)^2 + k_\varepsilon\right) |\nabla u(\omega, x)|^2 \ dx \, d\Pi \qquad (6.13)$$
$$E_{\text{phase}}^s[\phi] := \mathbb{E}(E_{\text{phase}}[\phi]) = \int_\Omega \int_D \nu \varepsilon |\nabla \phi(\omega, x)|^2 + \frac{\nu}{4\varepsilon} (1 - \phi(\omega, x))^2 \ dx \, d\Pi$$

and we define the stochastic energy as the sum of these, i.e.,

$$E_{AT}^s[u, \phi] = E_{\text{fid}}^s[u] + E_{\text{reg}}^s[u, \phi] + E_{\text{phase}}^s[\phi]. \qquad (6.14)$$

The Euler-Lagrange equations of the stochastic Ambrosio-Tortorelli energy are obtained from the first variation of the energies (6.13). Since these are the expected values of the classical energies (3.12), the computations are analogous. For example, for a test function, i.e., a random field, $v : \Omega \times D \to \mathbb{R}$, we get

$$
\begin{aligned}
\frac{d}{dt} E^s_{\text{fid}}[u + t\,v]\Big|_{t=0} &= \frac{d}{dt} \int_\Omega \int_D \Big(u(\omega, x) + t v(\omega, x) - u_0(\omega, x)\Big)^2 \, dx \, d\Pi \Big|_{t=0} \\
&= \int_\Omega \int_D 2\Big(u(\omega, x) - u_0(\omega, x)\Big) v(\omega, x) \, dx \, d\Pi \, .
\end{aligned}
\tag{6.15}
$$

With analogous computations for the remaining energy contributions, we arrive at the following system of SPDEs: we search for random fields $u, \phi \in L^2(\Omega) \otimes H^1(D)$ as the weak solutions of

$$
\begin{aligned}
-\text{div}\left(\mu(\phi(\omega, x)^2 + k_\varepsilon)\nabla u(\omega, x)\right) + u(\omega, x) &= u_0(\omega, x) \quad &\text{almost surely in } D, \\
-\varepsilon \Delta \phi(\omega, x) + \left(\frac{1}{4\varepsilon} + \frac{\mu}{2\nu}|\nabla u(\omega, x)|^2\right)\phi(\omega, x) &= \frac{1}{4\varepsilon} \quad &\text{almost surely in } D.
\end{aligned}
\tag{6.16}
$$

This system is analogous to the classical system (3.13) in which stochastic images replace the classical images. Both equations are SPDEs because the respective coefficients, $\phi(\omega, x)^2$ and $|\nabla u(\omega, x)|^2$, are random fields. Moreover, the right-hand side of the first equation, $u_0(\omega, x)$, is a stochastic input image, i.e., a random field. As for the deterministic model, we solve the system by alternatingly fixing ϕ or u, solving the respective equation for image or phase field in (6.16), and iterating until a fixed point is found.

6.3.1   Γ-CONVERGENCE OF THE STOCHASTIC MODEL

Before we turn to the discretization of the stochastic Ambrosio-Tortorelli model, let us emphasize that theoretical results from the classical model transform straightforwardly to the stochastic version. Ambrosio and Tortorelli [4] proved the Γ-convergence of their model toward the Mumford-Shah model. It is possible to extend this result to prove Γ-convergence of the stochastic Ambrosio-Tortorelli energy E^s_{AT} toward a stochastic Mumford-Shah energy, which we define as the expected value of the classical energy $\mathbb{E}(E_{MS})$.

For the formulation of the result, we use the stochastic analog of the space $\mathcal{D}_{h,n}$ from [4] sloppily denoted with $L^2(\Omega) \otimes \mathcal{D}_{h,n}$, which contains admissible functions for the energies. In the notation of [4], n is the space dimension and $h = 1/\sqrt{\varepsilon}$. Thus, letting the phase field scale ϵ tend to zero is equivalent to letting $h \to \infty$. We then are able to prove, cf. [76],

Theorem 6.1   *The stochastic Ambrosio-Tortorelli model E^s_{AT} Γ-converges to the stochastic Mumford-Shah model $\mathbb{E}(E_{MS})$ as $\varepsilon \to 0$. More precisely, let $(u_h, \phi_h) \in L^2(\Omega) \otimes \mathcal{D}_{h,n}$ be a sequence of pairs of random fields that converges to (u, ϕ) in $L^2(\Omega) \otimes \mathcal{D}_{h,n}$. Then we have*

$$
\mathbb{E}\left(E_{MS}[u(\omega, x), K(\omega)]\right) \leq \liminf_{h \to \infty} E^s_{AT}[u_h(\omega, x), \phi_h(\omega, x)]
$$

and for every pair of random fields $(u, \phi) \in L^2(\Omega) \otimes \mathcal{D}_{h,n}$ there exists a sequence $(u_h, \phi_h) \in \mathcal{D}_{h,n}$ converging to (u, ϕ) such that

$$\mathbb{E}\left(E_{MS}[u(\omega, x), K(\omega)]\right) \geq \limsup_{h \to \infty} E_{AT}^s[u_h(\omega, x), \phi_h(\omega, x)].$$

In both inequalities, the stochastic edge set $K(\omega)$ is defined accordingly as the discontinuity set of $u(\omega, x)$.

6.3.2   WEAK FORMULATION AND DISCRETIZATION

The system (6.16) contains two elliptic SPDEs, which are supposed to be interpreted in the weak sense. As shown before in Sections 4.6 and 6.1, we multiply the equations by a test function $v \in L^2(\Omega) \otimes H^1(D)$, integrate over Ω with respect to the corresponding probability measure, and integrate by parts over the physical domain D. Consequently, for the first equation in (6.16), we get

$$\int_\Omega \int_D \mu \left(\phi(\omega, x)^2 + k_\varepsilon\right) \nabla u(\omega, x) \cdot \nabla v(\omega, x) + u(\omega, x)v(\omega, x)\, dx\, d\Pi$$
$$= \int_\Omega \int_D u_0(\omega, x)v(\omega, x)\, dx\, d\Pi \quad (6.17)$$

and an analogous expression for the second equation of (6.16). Here we assume homogeneous Neumann BC for u and ϕ such that no boundary terms appear in the weak forms. For the existence of solutions of this SPDE, the constant k_ϵ is supposed to ensure the positivity of the diffusion coefficient $\mu(\phi^2 + k_\epsilon)$ almost surely, as in (6.5).

The weak system (6.17) is discretized by a substitution of the polynomial chaos expansion (5.2) of the image and the phase field. Analogous to the expositions for the linear diffusion filtering in Section 6.1, we get the fully discrete systems

$$\sum_{|\alpha|=0}^{p} \left(M^{\alpha,\beta} + L^{\alpha,\beta}\right) U_\alpha = \sum_{|\alpha|=0}^{p} M^{\alpha,\beta} (U_0)_\alpha$$
$$\sum_{|\alpha|=0}^{p} \left(\varepsilon S^{\alpha,\beta} + T^{\alpha,\beta}\right) \Phi_\alpha = A_\beta \quad (6.18)$$

for all $|\beta| = 0, \dots, p$. The $M \times M$ matrices $M^{\alpha,\beta}$, $L^{\alpha,\beta}$, $S^{\alpha,\beta}$, and $T^{\alpha,\beta}$ are blocks of the overall system matrix. The mass and stiffness matrix are defined as before (cf. (6.7) and (6.8))

$$\left(M^{\alpha,\beta}\right)_{i,j} = \mathbb{E}\left(\Psi^\alpha \Psi^\beta\right) \int_D P_i(x)\, P_j(x)\, dx,$$
$$\left(S^{\alpha,\beta}\right)_{i,j} = \mathbb{E}\left(\Psi^\alpha \Psi^\beta\right) \int_D \nabla P_i(x) \cdot \nabla P_j(x)\, dx, \quad (6.19)$$

and furthermore

$$\left(L^{\alpha,\beta}\right)_{i,j} = \sum_k \sum_\gamma \mathbb{E}\left(\Psi^\alpha \Psi^\beta \Psi^\gamma\right) \widetilde{(\phi^2)}^k_\gamma \int_D \nabla P_i(x) \cdot \nabla P_j(x) P_k(x)\, dx\,,$$

$$\left(T^{\alpha,\beta}\right)_{i,j} = \sum_k \sum_\gamma \mathbb{E}\left(\Psi^\alpha \Psi^\beta \Psi^\gamma\right) u^k_\gamma \int_D P_i(x) P_j(x) P_k(x)\, dx\,.$$

Here, $\widetilde{(\phi^2)}^k_\gamma$ denotes the coefficients of the polynomial chaos expansion of the Galerkin projection of ϕ^2 onto the image space. The right-hand side vector of the phase field equation in (6.18) is

$$A^i_\beta = \int_\Omega \Psi^\beta\, d\Pi \int_D \frac{1}{4\varepsilon} P_i\, dx = \begin{cases} \int_D \frac{1}{4\varepsilon} P_i\, dx & \text{if } |\beta| = 0\,, \\ 0 & \text{else}, \end{cases}$$

where we have used the fact that the stochastic basis functions are all odd, except the first one. Note that the expectations of the products of stochastic basis functions involved above are again the components of the lookup table $C_{\alpha\beta\gamma}$ that was introduced in Section 4.6. Moreover, the deterministic integrals can be precomputed as well, because they depend only on the choice of basis functions on the regular quadrilateral or hexahedral image grids. With these precomputations, the assembly of the system matrix can be accelerated significantly. Analogous to the classical FEM, the resulting systems of linear equations can be treated by an iterative solver such as the method of conjugate gradients [41]. The memory consumption, however, is enormous, because the matrix has N^2-times the storage requirement of the deterministic matrix, where N is the dimension of the polynomial chaos. Thus, we use again the GSD method for the solution to avoid generation of the full SFEM system matrix.

6.3.3   RESULTS

In the following section, we demonstrate the performance and advantages of the stochastic Ambrosio-Tortorelli model on two data sets. As for the random walker method, our first input image data set consists of $M = 5$ samples from the artificial "street sequence" [66]. The second data set consists of $M = 45$ image samples from US imaging of a structure in the forearm, acquired within two seconds. From the samples, we compute the polynomial chaos representation using $n = 10$ (US), respectively, $n = 4$ (street scene) random variables with the method described in Section 5.2. The images have a resolution of 100×100 pixels for the street sequence and 129×129 pixels for the US data set. We use a polynomial degree of $p = 3$, which leads to a polynomial chaos dimension of $N = 286$ (US) and $N = 35$ (street scene), respectively. For the reduction of the complexity by the GSD, we set $K = 6$. Furthermore, we use $\nu = 0.00075$ and $k_\varepsilon = 2.0h$ in all computations, where h is the grid spacing. To show the influence of the random variables, we compare the results from the US data with the classical deterministic results.

We conclude from our results that the stochastic method yields much more information than the deterministic method. The expected value of the stochastic method is comparable to the result of the classical method. However, stochastic information about such chaos coefficients, variance, etc., is the real benefit of the method. For example, the variance indicates the robustness of the detected edges. This information is not available in the classical model.

Street Image Data Set.   When treating successive frames of an image sequence as different realizations of a static scene, the gray value uncertainty of the input data is naturally high when close to the edges of moving objects. Thus, we expect the highest phase field variance in these regions. The results depicted in Figure 6.5 match these expectations. Indeed, in the region around the wheels of the car and around the right shoulder of the person, the edge detection is most influenced by the moving camera and with respect to the varying gray values between the samples at the edges. Also, around the edges in the background, the variance increases due to the moving camera. However, the stochastic method can detect the edges in the image properly.

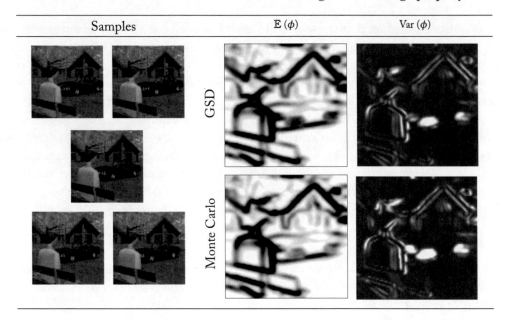

Figure 6.5: Segmentation result of the street scene. *On the left*, we show the five samples from which the stochastic input image is computed. *On the right*, we compare the results computed via the GSD method and a Monte Carlo sampling.

To verify the intrusive GSD method, we compared the results of the GSD implementation with Monte Carlo simulations using 10,000 samples. Figure 6.5 shows the results and reveals that both approaches produce similar results, but again, the GSD method is 100 times faster than performing Monte Carlo simulations with a moderate number of samples.

Ultrasound Samples.   The conversion of the input samples into the polynomial chaos, as described in Section 5.2, leads to the representation of the stochastic US image with 286 coeffi-

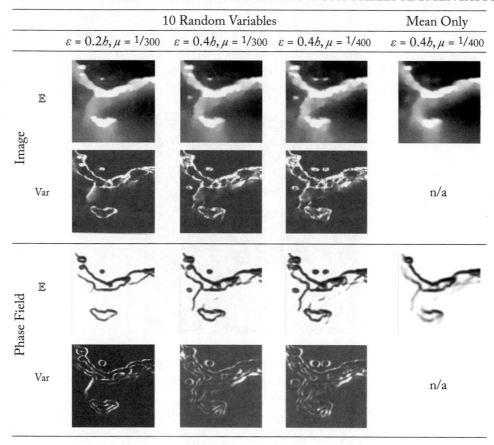

	10 Random Variables			Mean Only
	$\varepsilon = 0.2h, \mu = 1/300$	$\varepsilon = 0.4h, \mu = 1/300$	$\varepsilon = 0.4h, \mu = 1/400$	$\varepsilon = 0.4h, \mu = 1/400$

Figure 6.6: Mean and variance of the image and phase field for varying ε and μ using the US data. For comparison, we added the result from the deterministic method applied on the mean.

cients per pixel. Thus, a visualization of this stochastic image via stochastic moments, such as expected value and variance, is necessary. Figure 6.6 shows the expected value and the variance of the phase field ϕ and the smoothed image u for settings of the smoothing coefficient μ and the phase field width ε. The algorithm needs about 100 iterations, i.e., alternating solutions of (6.16) for u and ϕ, to compute a solution. From the variance image of the phase field, the identification of regions, where the input distribution has a strong influence on the segmentation result (areas with high variance), is straightforward. A benefit of the new stochastic edge detection via the phase field ϕ is that it allows for an identification of edges in a way that is robust with respect to parameter changes. In particular, within the four regions marked in Figure 6.7, the expectation of the phase field is highly influenced by the choice of μ and ν as shown in Figure 6.6. The blurred edge at position 1 can be seen in the expectation of the phase field only when we use a narrow phase field. In region 2, we have a different situation in which the edge can be identified using a wide phase field. In addition, the edges at positions 3 and 4 can be identified using adjusted parameters. However, note that one of these edges is not seen in the expectation of ϕ because of the particular choice of parameters; a high variance of ϕ indicates the possible existence of an edge, as in regions 1 and 2.

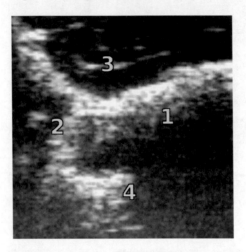

Figure 6.7: Mean value of the US data set for the stochastic Ambrosio-Tortorelli method with indication of image regions referred to in the text.

Moreover, the algorithm estimates the reliability of detected edges: A low expected value of the phase field and a low variance indicate that the edge is robust and not influenced by the noise and uncertainty of the acquisition process, which is true for the upper edges of the structure. In contrast, a high phase field variance, e.g., in the labeled regions 1–4, indicates regions where the detected edges are sensitive to noise. In addition, we can easily extract the distribution of the gray values for any pixel location inside the image and the phase field from the polynomial chaos expansion obtained via the GSD method.

6.4 STOCHASTIC LEVEL SET METHOD FOR IMAGE SEGMENTATION

As sketched in Chapter 3, level sets are widely used in applications ranging from computer vision [102] and material science to computer-aided design [94] for the tracking and representation of moving interfaces. When we try to combine a level set-based segmentation approach with stochastic images, we end up with a stochastic velocity for the level set propagation, i.e., we have to solve a hyperbolic SPDE. The development of numerical methods for hyperbolic SPDEs is an active research field. To date, there is no method available in the literature that can be applied to the stochastic level set equation. The use of classical methods, such as upwinding schemes [94], is not possible because they are based on the sign of the propagation speed, which, in the stochastic context, is a random variable, too. Thus, we use a parabolic approximation of the level set equation, which enables us to use the methods developed in the previous chapters.

Because of the importance of the level set equation in other applications besides the segmentation of images, this section is split into two parts. First, we present the derivation of the parabolic approximation of the stochastic level set equation along with the numerical discretization. Furthermore, we present numerical tests showing the applicability of the discretization. The second part of this section deals with the application of the stochastic level set equation for image segmentation. We introduce stochastic extensions of three widely used segmentation methods based on the level set equation: gradient-based segmentation, geodesic active contours, and Chan-Vese segmentation. The well-known deterministic versions of these models have been discussed in Chapter 3.

6.4.1   DERIVATION OF A STOCHASTIC LEVEL SET EQUATION

The discretization of the classical level set equation is based on techniques for the discretization of hyperbolic conservation laws. The discretization of hyperbolic SPDEs is a challenging task that is subject to ongoing research in many areas of application. We focus on a parabolic approximation of the level set equation to avoid the numerical problems related to the hyperbolic level set version, for which we have been inspired by the work of Sun and Beckermann for the classical level set equation [99].

Let us start by showing that the stochastic level set equation can be derived in a way that is similar to the deterministic case. The evolution of a zero level set of a random field level set function ϕ is characterized by

$$\phi(\omega, t, y_x(t, \omega)) = 0 \quad \text{almost surely in } \mathbb{R}_0^+ \times D, \tag{6.20}$$

where t is the time, ω a stochastic outcome, and $y_x(\omega, t)$ the (stochastic) path of a particle on the interface that passes through location $x \in D$. Obviously, for the following computations, we need to assume that the level set function is well defined, i.e., $|\nabla \phi(\omega, t, x)| \neq 0$ almost surely in $\mathbb{R}_0^+ \times D$. Taking the derivative w.r.t. t of this equation, and using the chain rule, we get the stochastic version of the advection equation

$$\phi_t(\omega, t, x) + F(\omega, t, x) \cdot \nabla \phi(\omega, t, x) = 0 \quad \text{almost surely in } \mathbb{R}_0^+ \times D, \tag{6.21}$$

where

$$F(\omega, t, x) = \frac{\partial y_x(\omega, t)}{\partial t}$$

is the velocity of the level set propagation. The velocity decomposes in a component in the normal direction N and in the tangential direction(s) T of the interface:

$$F(\omega, t, x) = F_N(\omega, t, x) + F_T(\omega, t, x), \tag{6.22}$$

where F_N and F_T are given by

$$
\begin{aligned}
F_N(\omega, t, x) &= (F(\omega, t, x) \cdot N(\omega, t, x))\, N(\omega, t, x), \\
F_T(\omega, t, x) &= F(\omega, t, x) - F_N(\omega, t, x).
\end{aligned}
\tag{6.23}
$$

Note that this decomposition depends on the stochastic outcome ω, because for every realization of the level set $\phi(\omega, \cdot, \cdot)$, we get a different normal $N(\omega, t, x) = |\nabla\phi(\omega, t, x)|^{-1}\nabla\phi(\omega, t, x)$ and a different decomposition of the stochastic quantity $F(\omega, t, x)$.

Substituting (6.22) and (6.23) into the advection equation (6.21) and using the relations

$$
\begin{aligned}
F_T(\omega, t, x) \cdot \nabla\phi(\omega, t, x) &= 0, \\
F_N(\omega, t, x) \cdot \nabla\phi(\omega, t, x) &= F_n(\omega, t, x)|\nabla\phi(\omega, t, x)|
\end{aligned}
\tag{6.24}
$$

yields the stochastic extension of the level set equation:

$$
\phi_t(\omega, t, x) + F_n(\omega, t, x)|\nabla\phi(\omega, t, x)| = 0.
\tag{6.25}
$$

In this equation, F_n is the speed of the level sets in their normal direction.

To avoid the use of a numerical upwinding scheme for hyperbolic SPDEs, we modify the stochastic level set equation in the spirit of Sun and Beckermann [99], see also Chapter 3. To this end, we further decompose the speed F_n into two components, one independent of and the other dependent on the interface curvature κ

$$
F_n(\omega, t, x) = a(\omega, t, x) - b(\omega, t, x)\kappa(\omega, t, x).
\tag{6.26}
$$

The curvature κ is expressed using the level set function ϕ, which is a standard approach for deterministic level sets [94]:

$$
\kappa(\omega, t, x) = \frac{1}{|\nabla\phi(\omega, t, x)|}\left(\Delta\phi(\omega, t, x) - \frac{\nabla\phi(\omega, t, x) \cdot \nabla(|\nabla\phi(\omega, t, x)|)}{|\nabla\phi(\omega, t, x)|}\right).
$$

This equation is valid for sufficiently smooth level set functions. If we prescribe a special behavior of the level set in its normal direction, some quantities such as the gradient or the curvature can be computed more easily. For example, we can set the level set function to

$$
\phi(\omega, t, x) = -\tanh\left(\frac{n(\omega, t, x)}{\sqrt{2}W}\right),
\tag{6.27}
$$

where n is the distance to the interface in the normal direction and $W \in \mathbb{R}$ an additional parameter controlling the width of the tangential profile. Then, we get a simple formula for the norm of the gradient:

$$
|\nabla\phi(\omega, t, x)| = -\frac{\partial\phi(\omega, t, x)}{\partial n} = \frac{1 - \phi(\omega, t, x)^2}{\sqrt{2}W}.
$$

Prescribing a special behavior of the level set function, such as the hyperbolic tangent profile, is a standard technique in the level set context. Another typical choice for classical level sets is to let ϕ be a signed distance function, i.e., to ensure that $|\nabla\phi| = 1$ (see [94]).

Collecting the above findings and substituting them into the level set equation (6.25) yields

$$\phi_t(\omega, t, x) + a(\omega, t, x) \frac{1 - \phi(\omega, t, x)^2}{\sqrt{2}W}$$
$$= b(\omega, t, x) \left(\Delta\phi(\omega, t, x) + \frac{\phi(\omega, t, x)\left(1 - \phi(\omega, t, x)^2\right)}{W^2} \right) \qquad (6.28)$$

almost surely in $\mathbb{R}_0^+ \times D$. This equation is parabolic for $b > 0$, and the hyperbolic term $a|\nabla\phi|$ is converted into a nonlinear term in ϕ. The resulting level set function will not be a signed distance function, but a phase field function with the hyperbolic tangent profile that we prescribed with (6.27).

In the absence of a curvature dependent speed, i.e., if $b = 0$ in (6.26), we can still retain the parabolic nature of the equation by subtracting curvature again from the right-hand side and keeping $b > 0$. This yields

$$\phi_t + a \frac{1 - \phi^2}{\sqrt{2}W} = b \left(\Delta\phi + \frac{\phi(1 - \phi^2)}{W^2} - |\nabla\phi| \mathrm{div} \left(\frac{\nabla\phi}{|\nabla\phi|} \right) \right), \qquad (6.29)$$

where we have omitted the arguments (ω, t, x) in the display. This subtraction is based on the idea of the counter term approach developed by Folch et al. [28] and should not be confounded with setting $b = 0$. In fact, the first term on the right-hand side conserves the tangential profile of the level set. If we set $b = 0$, the equation moves an arbitrary shaped level set instead of producing the tangential shaped level set. For further details, we refer to Chapter 3.

Despite the hyperbolic tangent profile that we needed for the above derivations, we can instead use a signed distance function again by using a nonlinear preconditioning technique [33]. Indeed, the substitution

$$\phi(\omega, t, x) = -\tanh\left(\frac{\psi(\omega, t, x)}{\sqrt{2}W} \right) \qquad (6.30)$$

ensures that ψ is a signed distance function to the interface because of (6.27) in which n is the distance to the interface. A stochastic signed distance function has to be a classical signed distance function for every realization $\omega \in \Omega$, as can be seen from the following theorem.

Theorem 6.2 *A stochastic signed distance function fulfills* $\mathbb{E}(|\nabla\phi|) = 1$ *and* $\mathrm{Var}(|\nabla\phi|) = 0$.

This is proved straightforwardly by noting that

$$\mathbb{E}(|\nabla\phi|) = \int_\Omega |\nabla\phi(\omega, t, x)| d\Pi = \int_\Omega 1 \, d\Pi = 1.$$

Also, we find that

$$\mathrm{Var}(|\nabla\phi|) = \int_\Omega (|\nabla\phi(\omega, t, x)| - 1)^2 d\Pi = \int_\Omega 0 \, d\Pi = 0.$$

$\square$

Substituting (6.30) into (6.29), we arrive at the final version of the stochastic level set equation:

$$\psi_t + a|\nabla\psi| = b\left(\Delta\psi + \frac{(1 - |\nabla\psi|^2)\sqrt{2}}{W}\tanh\left(\frac{\psi}{\sqrt{2W}}\right) - |\nabla\psi|\nabla\cdot\left(\frac{\nabla\psi}{|\nabla\psi|}\right)\right),$$

$$\text{a.s. in } \mathbb{R}_0^+ \times D, \tag{6.31}$$

where we omit displaying the dependence on (ω, t, x). The right-hand side of this level set equation serves as an integrated reinitialization scheme for the level set ψ. We use this form for some $b > 0$ when the speed from the decomposition (6.26) is not curvature dependent. When the speed does depend on the curvature, the last term of the right-hand side must be omitted.

6.4.2 INTERPRETATION OF STOCHASTIC LEVEL SETS

Equation (6.31) will produce time-dependent random fields, which we have to interpret as the evolution of level sets with random speed. Due to the random quantity that controls the speed of the level set motion, the position of the level sets is a random quantity, too.

One way to estimate the influence of the random speed component on the level set motion is to calculate the probability that the zero level set is at a specific position. Furthermore, we can calculate the whole band, in which we have positive probability that the zero level set is located there, i.e., where

$$\Pi\left(\phi(\omega, t, x) = 0\right) > 0 \tag{6.32}$$

holds. Also, in the normal direction of the expected value $\mathbb{E}(\phi) = 0$ of the zero level set location, we can estimate the PDF of the interface position (see Figure 6.8).

If we use Gaussian random variables, we end up with a nonzero probability for the interface location in the whole domain, which is due to the infinite support of Gaussian random variables. Thus, we limit our investigations to a polynomial chaos in uniform random variables. Uniform random variables have a compact support, leading to a band with finite thickness for the potential interface location.

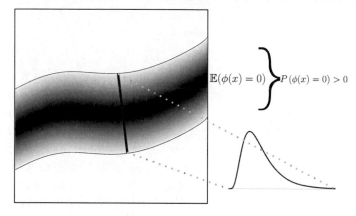

Figure 6.8: Stochastic level sets do not have a fixed position where $\phi(x) = 0$. Instead, there is a band with positive probability that the level set is equal to zero. Thus, it is possible to estimate the PDF of the interface location in the normal direction of the expected value of the interface (*lower right corner*).

6.4.3 DISCRETIZATION OF THE STOCHASTIC LEVEL SET EQUATION

We discretize (6.31) with an explicit Euler scheme for the time-derivative. Denoting with ψ^t the phase field at time t, we can write down the time discrete version of (6.31) as

$$\psi^{t+\tau} = \psi^t + \tau\left(-a|\nabla\psi^t| + b\left(\Delta\psi^t + \frac{\sqrt{2}}{W}\left(1 - |\nabla\psi^t|^2\right)\tanh\frac{\psi^t}{\sqrt{2}W} - |\nabla\psi^t|\mathrm{div}\left(\frac{\nabla\psi^t}{|\nabla\psi^t|}\right)\right)\right) \tag{6.33}$$

for all time-steps and almost surely in D. The spatial discretization is done by a uniform grid using finite differences or finite elements, cf. Chapters 2 and 4. The stochastic part is discretized using the polynomial chaos. Thus, we have to build numerical schemes in the polynomial chaos for the gradient norm, the curvature, and the hyperbolic tangent, which are used in (6.33).

Gradient Norm and Laplacian.   The gradient norm is obtained with finite difference approximations: The first-order directional derivatives are computed using central differences in the interior of the domain and forward respectively backward differences at the domain boundary. The necessary computations of the square and the square root are performed using the methods from Section 4.3. The Laplacian is evaluated as the sum of the second directional derivatives, which we compute using central differences in the interior and forward respectively backward differences at the boundary.

Hyperbolic Tangent.   We compute the hyperbolic tangent using the identity $\tanh(s) = 1 - 2(\exp(2s) + 1)^{-1}$. Thus, we use the computation of the exponential function in the polynomial chaos and other methods from Section 4.3. In other words, we compute the Galerkin projection of the hyperbolic tangent on the polynomial chaos.

Curvature. The computation of the curvature is the most critical process for the computation of the update, because the update is done in the whole domain, not just in a narrow band around the zero level set. Being the most critical process makes it necessary to compute a stable curvature even in regions with a high curvature. Such regions arise already in simple settings, e.g. , when the level set is initialized as a circle. The curvature in the midpoint of the circle goes to infinity. In principle, it is possible to use a simple finite difference approximation to get the mean curvature as the divergence of the normal. A more stable method for the curvature computation has been proposed by Sun and Beckermann [99] based on an idea by Echebarria et al. [24].

Let us finally note that due to the hyperbolic tangent profile of the level sets, we have to respect a condition on the maximal curvature of the represented object. For a high curvature, the hyperbolic tangent profiles would overlap for different points on the interface, which leads to instabilities of the numerical schemes for the discretization.

6.4.4 REINITIALIZATION OF STOCHASTIC LEVEL SETS

The right-hand side of (6.31) contains a reinitialization of the level set function. Following [99], this reinitialization yields accurate results for deterministic level sets. When using a stochastic velocity, however, we have to reinitialize all polynomial chaos coefficients, which have different scales. Typically, the first coefficient, the expected value, is orders of magnitude bigger than the remaining coefficients. Furthermore, the coefficients of polynomials in uncoupled random variables are close to zero. During our numerical experiments, we observed that the reinitialization via (6.31) is not sufficient. Thus, we need an additional reinitialization to get accurate stochastic results.

Classical reinitialization methods for level sets are not applicable in the stochastic context. The Fast Marching Method [94], for example, is based on an upwinding scheme for which stochastic variants are not available. Iterative reinitialization via $\phi_t = \text{sign}(\phi) \, (1 - |\nabla \phi|)$ is not possible because the signature of a stochastic quantity is not well-defined. The equation for energy minimization [57], which results from the observations of Theorem 6.2,

$$\alpha \left| \mathbb{E}(|\nabla \phi|) - 1 \right| + \beta \left| \text{Var}(|\nabla \phi|) \right| \to \min,$$

is unstable if ϕ converges to the stochastic signed distance function.

To get a working reinitialization scheme for stochastic level sets, we use a modification of the stochastic level set equation (6.31). As previously mentioned, the right-hand side of this function is an integrated reinitialization. We use this equation, set the speed to zero, i.e. , $a = 0$, and solve the equation for some artificial time T. Thus, our reinitialization equation is

$$\psi_t = b \left(\Delta\psi + \frac{(1 - |\nabla\psi|^2)\sqrt{2}}{W} \tanh\left(\frac{\psi}{\sqrt{2}W}\right) - |\nabla\psi|\text{div}\left(\frac{\nabla\psi}{|\nabla\psi|}\right) \right) \qquad \text{a.s. in } [0, T] \times D, \tag{6.34}$$

which we apply in all our numerical experiments.

6.4.5 NUMERICAL VERIFICATION

Before we apply the stochastic level set method to applications in image segmentation, we shall evaluate the proposed equation, its discretization and implementation. We are able to compare four implementations of the stochastic level set evolution: the intrusive implementation of the preconditioned phase field in the polynomial chaos (PC), a stochastic collocation approach based on the preconditioned phase field (SC), a Monte Carlo simulation of the preconditioned phase field (MC), and a Monte Carlo simulation of the original level set equation (MCL) $\phi_t + a|\nabla\phi| = 0$.

The comparison is performed on two typical tests for level set evolution: the evolution of a cosine curve in the inward and outward directions, and the evolution of an edge of a square in the inward and outward directions. Furthermore, we demonstrate the extension of the proposed method to three spatial dimensions on the Stanford bunny data set [103] and apply the preconditioned phase field equation with stochastic speed on it. In all numerical experiments, we set $W = 2.5h$, where h is the grid spacing. In the absence of a curvature dependent speed, we set $b = 1.25h$.

For the evolution of the cosine (see Figure 6.9), a challenge lies in the development of a shock [94] when the curve moves inward. To account for a stochastic velocity, we use a uniformly distributed speed $a \sim \mathcal{U}[1 - 0.2\sqrt{3}, 1 + 0.2\sqrt{3}]$. Consequently, the position of the shock is uncertain and the discretization has to be adequate in the vicinity of the possible shock positions. For the numerical experiments, we use a spatial resolution of 129×129 pixels, and a polynomial chaos in one random variable $n = 1$ with order $p = 2$, and we apply 30 time-steps with step size $\tau = 0.1h$. Furthermore, we computed 20 time-steps of the reinitialization equation (6.34) with step size $0.2h$ after each time-step. The polynomial chaos coefficients of the speed are set to $a_1 = 1$, $a_2 = 0.2$, and $a_3 = 0$, such that the expansion fulfills $\mathbb{E}(a) = 1$ and $\text{Var}(a) = 0.04$.

Figure 6.9 shows that all methods based on the stochastic preconditioned phase field formulation lead to the same results. Only the discretization of the deterministic level set method leads to different results, which is due to the reinitialization of the level set via fast marching [94]. The Fast Marching Method assumes that the value of the level set function at a grid point near the interface equals the signed distance to the interface. This assumption is not true in the presence of a shock, which leads to errors. For deterministic level sets, these errors do not matter, and also the expected value of stochastic level sets is accurate. However, for the stochastic part of the solution (the variance in Figure 6.9), which is orders of magnitude smaller than the expected

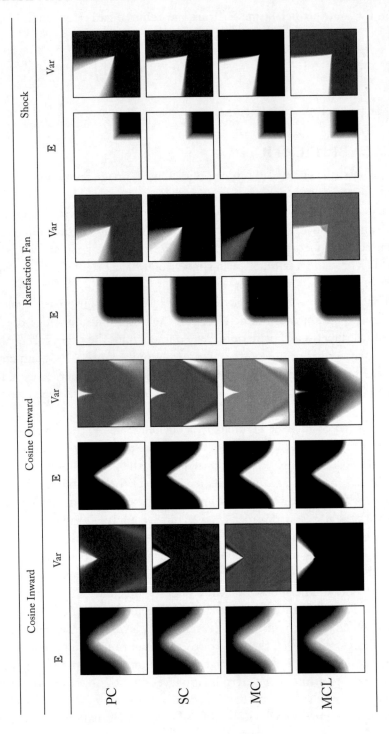

Figure 6.9: Comparison of expected value and variance of the resulting phase field for the cosine, the rarefaction fan, and the shock, classical tests for level set propagation of (6.31) using the polynomial chaos (PC), stochastic collocation (SC), Monte Carlo simulation (MC), and Monte Carlo simulation of the original level set equation (MCL).

value, the error becomes relevant. Thus, it is more precise to use the reinitialization via (6.34) in the presence of a shock.

When the interface moves outward, we obtain a rarefaction fan (see [94]). The same problem as for the shock arises, and again, the reinitialization via (6.34) is the better method.

For the second test, we let the edge of a square evolve in the inward and outward directions depicted in Figure 6.9. Again, we observe the development of a shock when the edge moves inward and of a rarefaction fan when the edge moves outward.

Our last test is the contraction of the Stanford bunny under uncertain velocity. Again, we used the speed $\mathbb{E}(a) = 1.0$ and $\mathrm{Var}(a) = 0.04$ from the previous tests. Figure 6.10 shows the results. For the Stanford bunny, we have the evolution of a 3D object. We use a method for the visualization of 3D stochastic images from Section 5.5 by visualizing the expected value color-coded by the variance. As expected, we see a high variance of the contour in regions with high curvature, which is due to the development of shocks, when the contour moves into the interior of these regions.

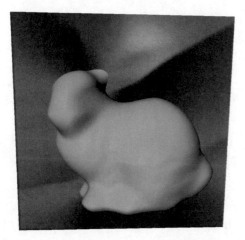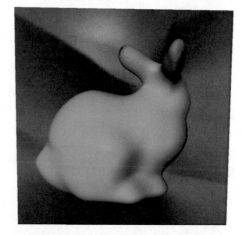

Figure 6.10: Expected value color-coded by the variance for the Stanford bunny after shrinkage under an uncertain speed in the normal direction. Red indicates regions with a high variance and green, regions with low variance. In addition, we show one slice of the variance.

6.4.6   IMAGE SEGMENTATION WITH STOCHASTIC LEVEL SETS

Let us now turn to the application of the stochastic level set equation to the segmentation of stochastic images. We discuss three stochastic segmentation methods that can all be formulated in the context of equation (6.31): gradient-based segmentation, geodesic active contours, and Chan-Vese segmentation. Other segmentation methods based on level sets can also be suitable for stochastic extensions through our stochastic level set equation.

Gradient-based Segmentation of Stochastic Images.   We replace the deterministic input image $u_0(x)$ with a stochastic image $u_0(\omega, x)$. The resulting equation for stochastic gradient-based segmentation is

$$\phi_t(\omega, t, x) + g[u_0](\omega, x)(1 - \varepsilon\kappa(\omega, t, x))|\nabla\phi(\omega, t, x)| = 0 \quad \text{almost surely in } \mathbb{R}_0^+ \times D \quad (6.35)$$

for some parameter $\varepsilon > 0$ and where the speed is given by the edge indicator

$$g[u_0](\omega, x) = (1 + |\nabla u_0(\omega, x)|^2)^{-1}. \quad (6.36)$$

This method can be implemented using the stochastic preconditioned phase field implementation, introduced in the last section, by rearranging the equation to $\phi_t(\omega, t, x) + F(\omega, t, x)|\nabla\phi(\omega, t, x)| = 0$ with the stochastic speed $F(\omega, t, x) = (1 - \varepsilon\kappa(\omega, t, x))(1 + |\nabla u_0(\omega, x)|^2)^{-1}$. Using the decomposition into curvature dependent and independent parts, we end up with

$$\phi_t + \frac{1}{1 + |\nabla u_0|}|\nabla\phi| = \frac{\varepsilon}{1 + |\nabla u_0|}\kappa|\nabla\phi| \quad \text{almost surely in } \mathbb{R}_0^+ \times D. \quad (6.37)$$

We can express this equation in the form of (6.31) by setting

$$a(\omega, t, x) := \frac{1}{1 + |\nabla u_0(\omega, x)|}, \quad b(\omega, t, x) := \frac{\varepsilon}{1 + |\nabla u_0(\omega, x)|}$$

and by omitting the counter term $|\nabla\psi|\text{div}(\frac{\nabla\psi}{|\nabla\psi|})$ that would be relevant in the case $\epsilon = 0$ only. Let us finally mention that it is straightforwardly possible to include a smoothing term in front of the gradient of the given image u_0.

Stochastic Geodesic Active Contours.   Geodesic active contours minimize the energy of a curve that is supposed to enclose an object in an image. For a stochastic curve $C(\omega, q) : \Omega \times [0, 1] \to D$ and a stochastic edge indicator $g[u_0](\omega, x) : \Omega \times \mathbb{R} \to \mathbb{R}$ on the given stochastic image u_0, the expected value of the geodesic curve energy is

$$\mathbb{E}(B(C)) = \int_\Omega \int_0^1 \beta g[u_0](C(\omega, q)) \, dq \, d\Pi + \int_\Omega \int_0^1 \alpha|C'(\omega, q)| dq \, d\Pi \quad (6.38)$$

for some user-prescribed parameters α, β. This energy minimizes the expected value of the curve's weighted length $\int_\Omega \int_0^1 |C'(\omega, q)| dq \, d\Pi$. It is minimal for short paths along edges inside the image. We define the edge indicator as the analog to the deterministic case as $g[u_0] = (1 + |\nabla G_\sigma * u_0|^p)^{-1}$, where G_σ is a Gaussian smoothing kernel with variance σ and $p \in \{1, 2\}$, cf. Section 3.5.

The stochastic Euler-Lagrange equation yields a necessary condition for a minimum. Their derivation is similar to the deterministic case [15, 53], when respecting the outer integration over

Ω. We end up with the stochastic equations analogous to (3.17):

$$\phi_t(\omega, t, x) = -\alpha \nabla g[u_0](\omega, t, x) \cdot \nabla \phi(\omega, t, x) + \beta g[u_0](\omega, t, x)|\nabla \phi(\omega, t, x)|$$
$$+ \varepsilon \kappa(\omega, t, x)|\nabla \phi(\omega, t, x)| \quad \text{almost surely in } \mathbb{R}_0^+ \times D. \tag{6.39}$$

In this equation, we have introduced an additional parameter ε to scale the influence of the curvature κ.

To use the stochastic preconditioned phase field equation (6.31), we have to split the velocity fully into normal and tangential components to arrive at the corresponding values for the random fields a and b. Again, the counter term in (6.31) must be omitted as we have a curvature dependent speed. We discretize the equation with the polynomial chaos and utilize computations and Galerkin projections in the polynomial chaos, cf. Section 4.3, accordingly to obtain a and b.

Stochastic Chan-Vese Segmentation.   We derive the stochastic Chan-Vese model from the classical Chan-Vese model by replacing all quantities by their stochastic counterparts. For user-defined parameters $\varepsilon, \mu, \lambda_1$ and λ_2, we have

$$\phi_t(\omega, t, x) = \delta_\varepsilon(\phi(\omega, t, x))\left(\mu \operatorname{div}\left(\frac{\nabla \phi(\omega, t, x)}{|\nabla \phi(\omega, t, x)|} \right) - v \right.$$

$$\left. - \lambda_1(u_0(\omega, x) - c_1(\omega, t))^2 + \lambda_2(u_0(\omega, x) - c_2(\omega, t))^2 \right), \tag{6.40}$$

where the phase field ϕ is a time-dependent random field and the initial image u_0 is a random field. The function δ_ε is the derivative of the stochastic smooth Heaviside approximation

$$H_\varepsilon(z(\omega)) = \frac{1}{2}\left(1 + \frac{2}{\pi} \arctan\left(\frac{z(\omega)}{\varepsilon} \right) \right)$$

for some random variable $z(\omega)$. Thus,

$$\delta_\varepsilon(z(\omega)) = \frac{1}{\pi \varepsilon + \frac{\pi}{\varepsilon} z(\omega)^2}.$$

The mean value of the object and the background become random variables through an averaging over the spatial dimensions, i.e., over the deterministic image domain only:

$$c_1[\phi](\omega, t) = \frac{\displaystyle\int_D u_0(\omega, x) H_\varepsilon(\phi(\omega, t, x))\, dx}{\displaystyle\int_D H_\varepsilon(\phi(\omega, t, x))\, dx}$$

$$c_2[\phi](\omega, t) = \frac{\displaystyle\int_D u_0(\omega, x)\,(1 - H_\varepsilon(\phi(\omega, t, x)))\, dx}{\displaystyle\int_D (1 - H_\varepsilon(\phi(\omega, t, x)))\, dx}. \tag{6.41}$$

In this equation, we have to evaluate the Heaviside approximation, which involves the computation of the inverse tangent of a stochastic quantity. To avoid the necessity to develop a numerical scheme for the stochastic inverse tangent, we use the approximation from [88]:

$$
\arctan(x) \approx
\begin{cases}
\dfrac{x}{1 + 0.28x^2} & \text{if } |x| \le 1, \\[2ex]
\dfrac{\pi}{2} - \dfrac{x}{x^2 + 0.28} & \text{else.}
\end{cases}
\tag{6.42}
$$

The remaining parts are discretized straightforwardly by the methods described before. As for the deterministic model (3.18), a solution of the full model is obtained by alternating between computing the polynomial chaos expansion of the random variables c_1 and c_2 for a fixed phase field ϕ and then solving (6.40) for these fixed c_1 and c_2.

Variance as Homogenization Criterion for Stochastic Chan-Vese Segmentation.   The main driving force of the stochastic Chan-Vese model is the difference between the mean value of the separated region and the actual gray value. In the polynomial chaos discretization, the mean value of the image regions is computed via an averaging of a collection of random variables. Thus, stochastic information from the input image made of no effect in the stochastic Chan-Vese model, because we are approximating the noise-free mean value when we average random variables.

Therefore, it is necessary to include further terms in the model to take stochastic information from the input image into account. Such terms can be fidelity terms for higher-order stochastic modes [101], e.g. , a homogenization of the variance. This extension can also be justified from the viewpoint of applications. In medical images, different organs or tissue components can have different noise levels. Thus, they can be separated by homogenizing the variance.

To include the homogenization of the variance, we introduce two further parameters ρ_1 and ρ_2 and consider

$$
\phi_t = \delta_\varepsilon(\phi)\left(\mu \operatorname{div}\left(\frac{\nabla\phi}{|\nabla\phi|} \right) - v - \lambda_1(u_0 - c_1)^2 + \lambda_2(u_0 - c_2)^2 - \right.
$$
$$
\left. \rho_1\left(\operatorname{Var}(u_0) - v_1\right)^2 + \rho_2\left(\operatorname{Var}(u_0) - v_2\right)^2 \right), \tag{6.43}
$$

where we do not denote the dependencies on ω, t, and x. Analogous to (6.41), the random variables v_1 and v_2 are defined as

$$v_1[\phi](\omega,t) = \frac{\int_D \text{Var}(u_0(\omega,x))H_\varepsilon(\phi(\omega,t,x))\,dx}{\int_D H_\varepsilon(\phi(\omega,t,x))\,dx}$$

$$v_2[\phi](\omega,t) = \frac{\int_D \text{Var}(u_0(\omega,x))(1 - H_\varepsilon(\phi(\omega,t,x)))\,dx}{\int_D (1 - H_\varepsilon(\phi(\omega,t,x)))\,dx}.$$

Also, it is possible to homogenize every polynomial chaos coefficient independently, leading to various additional constraints.

6.4.7 RESULTS

For the presentation of the results of the level set-based segmentation of stochastic images, we use two data sets (see left column of Figure 6.11). The first data set consists of 289 reconstructions of a CT data set showing a slice through a human head with 100×100 pixels, see Section 5.2. These reconstructions are treated as independent realizations of a stochastic image, for which the polynomial chaos expansion is calculated using the methods from Section 5.2. The second data set consists of a liver mask embedded into a 129×129 pixel image with a varying gradient strength in the background. This image is corrupted by uniform noise, and 25 samples, i.e., noise realizations, of this image are treated as input for the generation of the stochastic image. In both data sets, the generated stochastic image contains two random variables, and we use a polynomial degree of two, i.e., $n = 2$ and $p = 2$.

Stochastic Gradient-based Segmentation.   For the gradient-based segmentation, we set the parameter ε to the grid spacing h and the level set is initialized as a circle in the center of the image with a radius of 0.15 of the image width. Figure 6.11 shows information generated during the level set evolution. Firstly, some realizations of the contour after 280 iterations are shown on the expected value of the input image. The images show that the noise in the input image influences the segmentation in regions with a low gradient, i.e., in the bone regions of the head phantom. In regions where the level set has not entered the bone and in regions where the evolution has reached the outer bone boundary, the segmentation is more stable with respect to noise. The stability can be inferred by examining the realizations of the contour lines, which are close together in these regions. Secondly, in Figure 6.11 the variance of the level set function after 280 iterations is shown. From this image, the variance is clearly visible as locally constant in directions of the normal to the contour. Thirdly, we show some "expected contours," i.e., zero-level set contours of the expected value of the level set function, during the evolution. The contours show the typical behavior of a rapid propagation of the level set toward

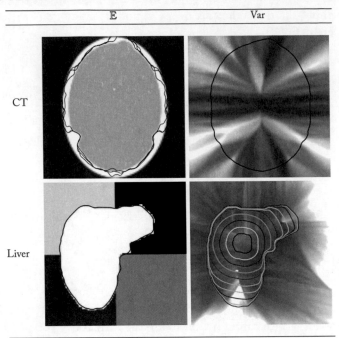

Figure 6.11: Results from the stochastic gradient based segmentation. (*top left*) Some realizations of the contour after 280 iterations are shown on the expected value of the input image. (*top right*) The expected contour after 280 iterations is shown on the variance of the level set function. (*bottom left*) The expected contour after 280 iterations is shown on the expected value of the input image. (*bottom right*) Some expected contours during the level set evolution are shown on the variance of the level set after 280 iterations.

the object boundary and the influence of the stopping function that tries to stop the evolution at the boundary.

Stochastic Geodesic Active Contours.   We show similar information for the stochastic geodesic active countour method shown in Figure 6.12. The final time has been reached after 240 iterations of time-step size $0.2h$. For the liver data set, we see that as for the gradient-based segmentation, the regions with high variance are at the bottom and the upper-right side of the object. The advantage of the stochastic geodesic active contour approach over the stochastic gradient-based segmentation is that running over the edges is mostly avoided.

Stochastic Chan-Vese Segmentation.   In Figure 6.13, we show some results of the segmentation with the stochastic Chan-Vese approach. For the head data set, we see that the model slightly overestimates the object, because the final expected contour is not perfectly aligned with the boundary. This overestimation is due to the homogenization criterion that the stochastic

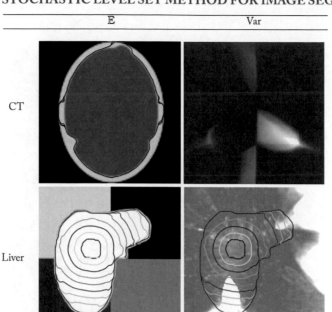

Figure 6.12: Results from the segmentation with stochastic geodesic active contours. (*top left*) Some realizations of the contour after 240 iterations are shown on the expected value of the input image. (*top right*) Variance of the level set function after 240 iterations. (*bottom*) Some expected contours during the level set evolution are shown on the expected value of the input (*bottom left*) and the variance of the level set function (*bottom right*) after 240 iterations.

Chan-Vese model tries to fulfill. Also, we see from the variance image that the segmentation is uncertain in the critical areas at the bottom and top. Furthermore, the variance identifies two critical regions on the right and left of the head object.

Figure 6.14 shows a comparison of realizations of the final contours for the three stochastic level set based segmentation methods discussed here. We see that the gradient-based segmentation is the most sensitive to the noise, whereas the geodesic active contours and the Chan-Vese segmentation are more robust.

For the liver data set, we show the level set evolution on the expected value of the initial image and on the variance of the final level set in Figure 6.13. Again, the variance image indicates that there is some uncertainty in the final contour. However, from Figure 6.15, we see that realizations drawn from the final stochastic level set are very close to each other and to the final contour; thus, there is very small variance in the final segmentation result only. This behavior results from the way in which this image was constructed: We have added artificial noise to a noise-free image. This noise nearly cancels out due to the averaging process for the computa-

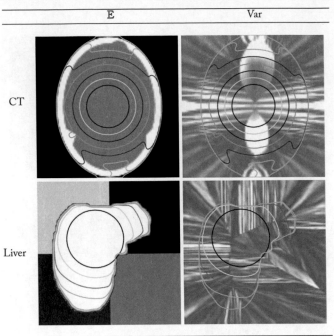

Figure 6.13: Results from the segmentation with the stochastic Chan-Vese approach. The expectations of the input image, the variance of the phase field, and several expected contours during the evolution are shown.

tion of the random variable for the mean inside the regions. Thus, the final result is (almost) deterministic.

The extension of the stochastic Chan-Vese approach that homogenizes the variance of the object and the background allows the segmentation of objects in images with a constant mean. With this method, an object can be differentiated from the background by just the variance of the input even if the object is not visible in the mean. In such a case, a classical method on the mean image would fail. Figure 6.16 shows the result of the segmentation of an image with a constant mean but nonconstant variance. In samples, the object is visible because of the different noise levels, but, again, the classical Chan-Vese approach cannot segment the object. The variance extension of the stochastic Chan-Vese approach yields the correct result.

6.5 VARIATIONAL METHOD FOR ELASTIC REGISTRATION

From the multitude of possible registration approaches, we focus on the stochastic variant of elastic registration. From the deterministic version (3.30), we arrive at the stochastic variant by

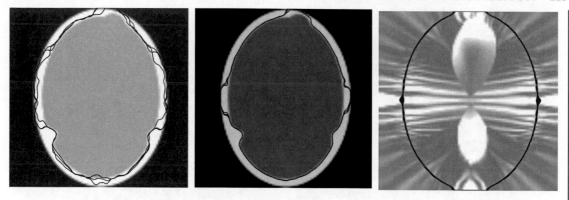

Figure 6.14: Some realizations of the final contours from gradient-based segmentation (*left*), geodesic active contours (*middle*) and Chan-Vese segmentation (*right*) are shown. For the Chan-Vese segmentation we show the variance image of the final phase field.

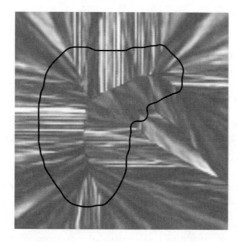

Figure 6.15: Some realizations of the final contour from the stochastic Chan-Vese segmentation of the liver data set are shown on the variance image of the final phase field (cf. Figure 6.13). The realizations lie so close together that no difference can be perceived. Thus, variance in the final segmentation result is present on very small scales only.

straightforward replacement of quantities with random fields. Let $u \in L^2(\Omega) \otimes H^1(D)$ be the stochastic reference image, and $v \in L^2(\Omega) \otimes H^1(D)$ be the stochastic template image. We are looking for a stochastic deformation vector field $\mathbf{w} \in L^2(\Omega) \otimes \left(H^1(D)\right)^d$ that minimizes the

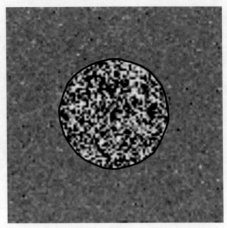

Figure 6.16: Variance of the stochastic image to segment (*left*). The expected value is not depicted, because the expected value is an image with the same gray value at every pixel. On the right, the segmentation result is depicted on one realization (one sample) of the stochastic image to segment.

stochastic elastic energy,

$$
\begin{aligned}
E_{ER}^s[\mathbf{w}] &= \mathbb{E}(E_{ER}[\mathbf{w}]) \\
&= \frac{1}{2} \int_\Omega \int_D |v_{\mathbf{w}}(\omega, x) - u(\omega, x)|^2 \, dx \, d\Pi \\
&\quad + \frac{\mu}{4} \int_\Omega \int_D \left| \nabla \mathbf{w}(\omega, x) + (\nabla \mathbf{w}(\omega, x))^T \right|^2 \, dx \, d\Pi \\
&\quad + \frac{\lambda}{4} \int_\Omega \int_D |\mathrm{div}(\mathbf{w}(\omega, x))|^2 \, dx \, d\Pi \,,
\end{aligned}
$$

for some user-defined parameters λ and μ, and where $v_{\mathbf{w}}(\omega, x) = v(\omega, x - \mathbf{w}(\omega, x))$ is the deformed template image as before. Let us emphasize that the stochastic generalizations of other distance measures, such as mutual information [106] or normalized gradient fields [69], could be used as well and would not alter the derivation of the system matrix. In fact, the distance measure influences the right-hand side of the system only.

We proceed with the derivation of the Euler-Lagrange equations, which is done analogously to the deterministic case but with additionally respecting the integration over Ω. Consequently, we arrive at the stochastic Navier-Lamé equation,

$$
\mu \Delta \mathbf{w}(\omega, x) + (\lambda + \mu) \nabla \mathrm{div}(\mathbf{w}(\omega, x)) = F[u, v, \mathbf{w}](\omega, x) \quad \text{a.s. in } D \tag{6.44}
$$

with right-hand side $F[u, v, \mathbf{w}](\omega, x) = (v_{\mathbf{w}}(\omega, x) - u(\omega, x)) \nabla v_{\mathbf{w}}(\omega, x)$. Note that this is a vector-valued equation.

We transform to the weak form and discretize it with the polynomial chaos as before. Furthermore, we denote the vector of unknowns with $(W_1, \ldots, W_d)$, where the component

$W_m \in \mathbb{R}^{NM}$ has N stochastic modes in $\mathbb{R}^M$. This process yields the linear block system of equations,

$$\begin{pmatrix} \mu S + (\lambda + \mu)R_{11} & \cdots & (\lambda + \mu)R_{1d} \\ \vdots & & \vdots \\ (\lambda + \mu)R_{d1} & \cdots & \mu S + (\lambda + \mu)R_{dd} \end{pmatrix} \begin{pmatrix} W_1 \\ \vdots \\ W_d \end{pmatrix} = \begin{pmatrix} F_1 \\ \vdots \\ F_d \end{pmatrix}, \qquad (6.45)$$

where S is the block-matrix from (6.8) and where the blocks $R_{mn} = (R_{mn}^{\alpha,\beta})_{|\alpha|,|\beta|=0,\dots,p}$ are defined as

$$(R_{m,n}^{\alpha,\beta})_{i,j} = \mathbb{E}(\Psi^\alpha \Psi^\beta) \int_D \partial_m P_i(x)\partial_n P_j(x)\,dx.$$

The right-hand side contains vectors $F_m \in \mathbb{R}^{NM}$ with

$$(F_m^\beta)_j = \sum_{i=1}^M \sum_{k=1}^M \sum_{|\alpha|=0}^p \sum_{|\gamma|=0}^p \left((v_w)_\alpha^i - u_\alpha^i\right)(v_w)_\gamma^k \, \mathbb{E}\left(\Psi^\alpha \Psi^\beta \Psi^\gamma\right) \int_D P_i(x)\partial_m P_k(x)P_j(x)\,dx$$

for $|\beta| = 0, \dots, p$ and $j = 1, \dots, M$ and with the notation of the polynomial chaos coefficients as before.

6.5.1   RESULTS

For the numerical experiments, we used a 65×65 pixel resampled version of a sequence from the optical flow evaluation data set [66]. To generate a stochastic input from this deterministic test sequence, we used three consecutive frames and treated them as samples from an uncertain acquisition process. These samples have been mapped to the polynomial chaos with $n = 1$ and $p = 1$ for the calculations. The mean and variance of this mapping are depicted in Figure 6.17. As expected, the images show a high uncertainty around the edges of the moving objects in the image. For the Navier-Lamé equation, we used the deterministic parameters $\mu = 0.1$ and $\lambda = 0.01$ to control the influence of the diffusion and deformation part of the equation. The results after iterating the stochastic system to a steady state are depicted in Figure 6.18. As expected, the uncertainty/variance is high around the corners of the moving object in the direction of motion. Note that the variance of the y component of the deformation field has been multiplied by a factor of 10 compared to the variance of the x component to be shown with the same visualization parameters.

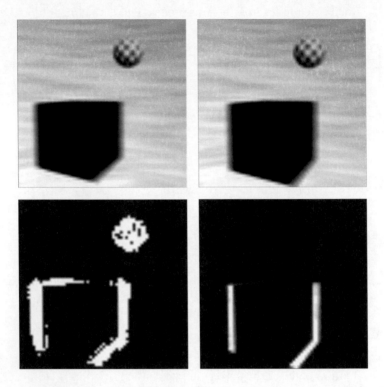

Figure 6.17: Mean (*top row*) and variance (*bottom row*) of the stochastic reference (*left column*) and template (*right column*) image.

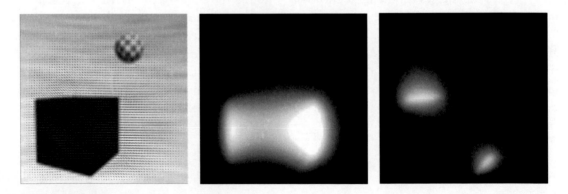

Figure 6.18: Visualization of the mean of the deformation field (*left*) and variance of the vector field components (*middle:* x-component, *right:* y-component). Note that the variance of the y component has been scaled by a factor of 10.

CHAPTER 7

Sensitivity Analysis

In the previous chapters, we have discussed methods for the denoising, segmentation, and registration of stochastic images. Starting from a stochastic input image, these methods yield a stochastic image as the result. For image segmentation, this process permits the characterization of the influence of gray value uncertainty in the input data on the segmentation result.

This chapter discusses another use for stochastic PDEs in computer vision, the analysis of the sensitivity of computer vision models to user-prescribed parameters [11, 91]. Studying the influence of parameters is not new to the image processing community [9, 98, 117], but such a study is often left out for performance reasons. Classically, a sensitivity analysis is done by a Monte Carlo simulation with respect to the assumed distribution (often uniform) of the model parameters. Such sampling is time-consuming because the underlying image processing PDE must be solved for every Monte Carlo sample. Due to the slow convergence rate of the Monte Carlo method, thousands of samples are needed to get accurate results. Another possibility is to perform a sensitivity analysis based on Bayesian inference [74].

Our method avoids sampling from the parameter distribution. Instead, we solve a dedicated stochastic PDE at the cost of a few Monte Carlo samples. In contrast to the concepts presented so far in this book, we do not work on stochastic images as input data. Instead, the central idea lies in replacing the deterministic parameters by random variables and applying the methods to deterministic images. Still, the outputs will be stochastic images, which are comparable to the results of the methods from Chapter 6. The difference is that the results are stochastic due to the stochastic parameters instead of a stochastic input image. We visualize the results using the same techniques as for stochastic images, showing the influence of the parameters on the segmentation result.

When focusing on segmentation with our approach, we detect regions in the image that are highly influenced by the choice of the segmentation parameters and regions, where the segmentation is robust with respect to parameter changes. In addition, we can investigate which segmentation parameters have a strong influence on the segmentation result. For example, with geodesic active contours, the influence of the smoothing term should be nearly the same on the whole image, whereas the weight related to the edge detector is important on the edges in the image. Our approach needs only a few random variables, typically one for every segmentation parameter. Hence, it is suitable for a straightforward discretization with the polynomial chaos without the need to reduce the number of random variables via the Karhunen-Loève decomposition.

We demonstrate the performance of our concept by applying the sensitivity analysis to some famous PDE-based image processing operators: Perona–Malik diffusion, random walker segmentation, discontinuity preserving optical flow, and Ambrosio-Tortorelli segmentation, as well as gradient-based segmentation and geodesic active contours.

7.1   CLASSICAL SENSITIVITY ANALYSIS

The investigation of the sensitivity of image processing operators with respect to parameter changes has found some attention in the image processing community [9, 98, 117]; however, the existing approaches are rarely used due to their high computational costs. The goal of classical sensitivity analysis is to identify the "important" parameters, i.e., the parameters with the strongest influence on the result, as well as parameter ranges where the result is sensitive to small variations of these parameters.

If the sensitivity analysis is based on Monte Carlo simulation, the procedure is the following: Based on the assumed distribution of the model parameters under investigation, R samples of the parameters are generated. The image processing task is then performed independently for every sample parameter, leading to realizations $u^{(1)}, \ldots, u^{(R)}$. Afterward, stochastic information is generated from the sample results by the well-known approximating formulas for mean and variance:

$$\mathbb{E}(u) \approx \frac{1}{R} \sum_{i=1}^{R} u^{(i)}, \qquad \text{Var}(u) \approx \frac{1}{R-1} \sum_{i=1}^{R} \left(u^{(i)} - \mathbb{E}(u) \right)^2 . \qquad (7.1)$$

To get accurate results, a huge number of samples are needed because of the slow convergence of the samples' mean and variance toward the real values. Remember, the approximation is of the order $\mathcal{O}\left(1/\sqrt{R}\right)$. As an example of a classical sensitivity analysis, we show segmentation results obtained from the random walker approach with a slightly varying parameter in Figure 7.1.

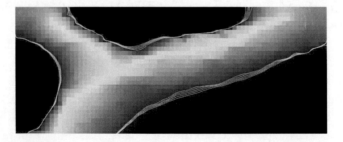

Figure 7.1: Various segmentation results obtained via a random walker segmentation of a medical ultrasound image of a structure in the forearm with a slightly varying parameter.

7.2   PERONA-MALIK DIFFUSION

Perona-Malik diffusion [81] smooths an image and preserves edges during the smoothing process. In the PDE (3.4), the coefficient g is an edge indicator that is usually defined as $g(s) = (1 + s^2/\lambda^2)^{-1}$ for some $\lambda > 0$. For the sensitivity analysis, we identify the parameter λ with a random variable that we discretize with the polynomial chaos, thus,

$$\lambda(\xi) = \sum_{|\alpha|=0}^{p} \lambda_\alpha \Psi^\alpha(\xi) \tag{7.2}$$

with polynomial chaos coefficients $\lambda_\alpha \in \mathbb{R}$. Consequently, the edge indicator becomes a random variable

$$g(\xi, s) = \frac{1}{1 + s^2/\lambda(\xi)^2}, \tag{7.3}$$

and the diffusion coefficient $g(\xi, |\nabla u_\sigma|)$ becomes a random field.

Note that here and in the following section, we directly consider the parameter random variables to be defined on the stochastic space Γ, thus they depend on the random variables ξ. The reason for this dependence is that the parameters become random variables that we design artificially. Thus, we do not base them on the unknown sample space Ω but on the well-known space Γ, thereby sparing the transformation from Ω to Γ. To ease the notation, we also consider all other random fields directly on $\Gamma \times D$, keeping in mind that this expression actually means $\xi = \xi(\omega)$, cf. Section 4.3.

As a consequence of the stochastic parameter, we obtain a stochastic version of the Perona-Malik equation

$$u_t(\xi, x) - \text{div}\,(g(\xi, |\nabla u_\sigma(\xi, x)|)\nabla u(\xi, x)) = 0 \qquad \text{almost surely in } \mathbb{R}_0^+ \times D,$$
$$u(\xi, x) = u_0(x) \quad \text{almost surely in } D.$$

Note that although we use a deterministic input image u_0, the solution of the equation will be stochastic because of the stochastic diffusion coefficient.

Discretizing u with the polynomial chaos as in Section 6.1 and using an explicit Euler scheme with time-step τ for the discretization of the time-derivative, we arrive at the same block system as before, i.e., (6.6). Again, the stiffness matrix is given by $L^{\alpha,\beta}$ from (6.7). For the computation of the coefficients g_γ^k, we straightforwardly use the rules for calculations described in Section 4.3.1. In a final step, we do a Galerkin projection of the resulting formula on the polynomial chaos to obtain the relevant coefficients.

Again, the smoothing of the stochastic image is obtained by one time-step of length 0.5σ of the stochastic heat equation before the calculation of the edge indicator, see Section 6.1. Note that we could also study the influence of the pre-smoothing parameter σ on the edge indicator and thus the final smoothing result. Since we are utilizing the equivalence of Gaussian smoothing with solving the heat equation, we are considering a stochastic stopping time

$T(\xi) = 0.5\sigma(\xi)^2$. However, it is also possible to keep the stopping time fixed at $T = 1$ and use a stochastic diffusion coefficient $0.5\sigma(\xi)^2$ instead.

To demonstrate the sensitivity analysis of Perona-Malik diffusion, we use a test image with 187×187 pixels, corrupted by uniform noise (see Figure 7.2). We performed 250 steps of Perona-Malik diffusion with time-step $\tau = \frac{1}{187 \cdot 187 \cdot 6}$. The parameter $\lambda \sim \mathcal{U}[0.16, 0.24]$ is uniformly distributed, varying 20% around its mean. We model this random variable by a Legendre gPC expansion with coefficients $\lambda_1 = 0.2$ and $\lambda_2 = 0.04$. Such uniform distribution is the best choice when only lower and upper bounds for the parameter are known. To capture the nonlinear effects influencing the stochastic result, we use a gPC in one random variable with polynomial degree $p = 5$ to model the output image u.

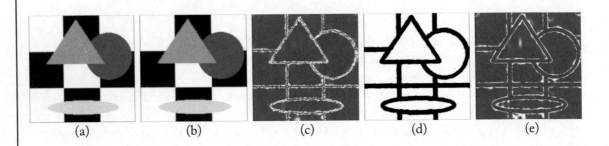

(a)   (b)   (c)   (d)   (e)

Figure 7.2: Initial image (a), mean (b), and variance (c) for Perona-Malik smoothing of an image corrupted by uniform noise; mean (d), and variance (e) of the stochastic edge indicator.

The results of this experiment are depicted in Figure 7.2. The Perona-Malik diffusion smoothes the noise in the input image. However, in the vicinity of the edges, there is high uncertainty whether the region belongs to the inner part of the objects or to the background. This uncertainty about the location of the objects is also visible in the edge indicator of the smoothed image. In some parts of the image, there is a wide region around the true edge position where the mean value of the edge indicator is below 1. Furthermore, the high variance of the edge indicator identifies regions with a high uncertainty in the edge position.

In Figure 7.3, the Perona-Malik diffusion on a medical image is used to decompose the image into regions. As indicated by the edge indicator's variance and the variance of the final image, there is uncertainty in the decomposition of the image into the regions with respect to the parameter λ. Choosing different realizations of the parameter λ, we get different decompositions of the image (see Figure 7.4).

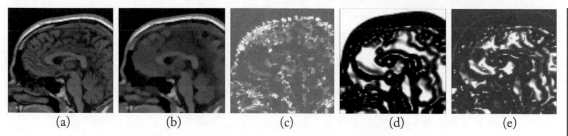

Figure 7.3: Initial image (a), expected value (b), and variance (c) for Perona-Malik smoothing of an MR image and expected value (d), and variance (e) of the edge indicator.

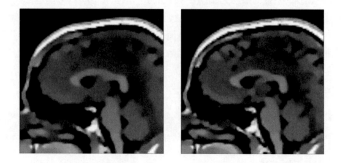

Figure 7.4: Two realizations of the stochastic Perona-Malik result. Depending on the choice of the parameter λ, we obtain different decompositions of the image.

7.3   RANDOM WALKER AND DIFFUSION IMAGE SEGMENTATION

According to the derivations in Section 3.3, the PDE for random walker segmentation [36] is given by

$$-\mathrm{div}(w\nabla u) = 0 \text{ in } D, \ u = 1 \text{ on } V_O, \ u = 0 \text{ on } V_B .$$

The weight w is defined as in (3.9), involving a user-defined parameter β. Identifying this parameter β with a random variable that is approximated in the polynomial chaos, i.e.,

$$\beta(\xi) = \sum_{|\alpha|=0}^{p} \beta_\alpha \Psi^\alpha(\xi),$$

we end up with a stochastic weight

$$w(\xi, x) = \exp\left(-\beta(\xi)\frac{|\nabla u_0(x)|^2}{\mathrm{supp}_D |\nabla u_0|^2}\right) = \sum_{i\in\mathcal{I}} \sum_{|\alpha|=0}^{p} w_\alpha^i \Psi^\alpha(\xi) P_i(x).$$

Consequently, the stochastic segmentation u is the solution to the SPDE

$$-\mathrm{div}(w(\xi, x)\nabla u(\xi, x)) = 0 \text{ almost surely in } D,$$
$$u(\xi, x) = 1 \text{ almost surely on } V_O,$$
$$u(\xi, x) = 0 \text{ almost surely on } V_B.$$

Proceeding with the polynomial chaos discretization, we obtain the linear system (6.11), where the matrices L_U and B are assembled with the new weights w_α^i.

We apply the sensitivity analysis of the random walker segmentation, first on a test image with resolution 129×129 pixels and using a uniformly distributed parameter $\beta \sim \mathcal{U}[8, 12]$. Thus, the parameter's expected value is 10 and its gPC coefficients are $\beta_1 = 10$ and $\beta_2 = 2$. Again, we used a polynomial chaos in one random variable and a polynomial degree of $p = 5$. The test image shows a liver mask in front of a varying background. We purposely smoothed the image with a Gaussian to end up with a blurry region, without a sharp gradient between object and background. The result of the random walker segmentation on this image with stochastic segmentation parameter β is shown in Figure 7.5. It is clear that the volume of the segmented object depends on the segmentation parameter.

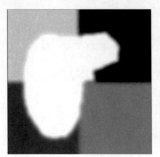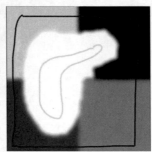

Figure 7.5: The blurry liver image (*left*), the seed regions (*middle*), and the contours obtained from the stochastic segmentation result (*right*).

Applying the sensitivity analysis on a medical ultrasound image with resolution 300×300 pixels, we get a stochastic segmentation result from which we visualize the mean and variance of the probability map in Figure 7.6. The variance image shows a high uncertainty in regions with a low gradient between object and background and in regions where the selection of the seed points missed important parts of the object. For example, at the upper boundary of the object in the upper right corner of the image, there is a high-intensity region inside the object that was not marked as a seed region. Due to the missing information in the algorithm, whether this region belongs to the object or to the background, the uncertainty has been increased.

To get a more intuitive interpretation of the stochastic segmentation result, we depicted realizations in Figure 7.7. The contours easily identify regions where the object boundary is highly influenced by the choice of the parameter β. Note that there is no one-to-one correspondence

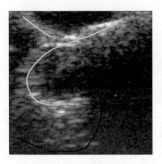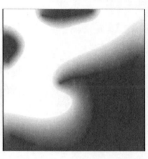

Figure 7.6: Seeds for object and background (*left*). Expected value (*middle*) and variance (*right*) of the probability map as a result of stochastic random walker segmentation.

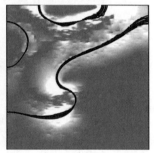

Figure 7.7: Samples of the stochastic contour encoded in the stochastic probability map shown on its mean (*left*) and variance (*right*).

between a high variance in the probability map and a high uncertainty in the contour positions, due to the varying gradient of the probability map. For more details, we refer the reader to [79].

Another observation from the stochastic segmentation result is that the PDF of the segmented volume (cf. Section 6.2.1) is not uniformly distributed, even though the stochastic parameter has a uniform distribution. Figure 7.8 shows the PDF of the segmented areas for both test examples. For the US data set, the resulting PDF is close to a uniform distribution. For the liver data set, the PDF concentrates around a peak. Both PDFs are computed using the method described in Section 6.2.

7.4   AMBROSIO–TORTORELLI SEGMENTATION

Sensitivity analysis for Ambrosio-Tortorelli segmentation requires the solution of a system of two coupled SPDEs and involves four random variables $\mu(\xi), \nu(\xi), \epsilon(\xi)$ and $k_\varepsilon(\xi)$ as user-defined parameters. As before, the stochastic parameters will render the result of the segmenta-

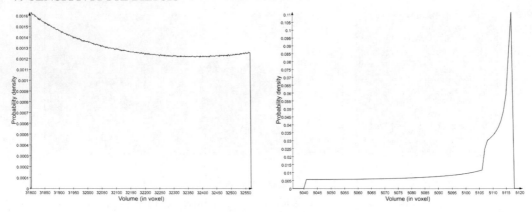

Figure 7.8: PDF of the area of the stochastically segmented objects for the ultrasound image (*left*) and the liver image (*right*).

tion stochastic. Thus, we are looking for solutions of the SPDE

$$-\mathrm{div}\Big(\mu(\xi)(\phi(\xi,x)^2 + k_\varepsilon(\xi))\nabla u(\xi,x)\Big) + u(\xi,x) = u_0(x) \quad \text{almost surely in } D,$$

$$-\epsilon(\xi)\Delta\phi(\xi,x) + \left(\frac{1}{4\epsilon(\xi)} + \frac{\mu(\xi)}{2\nu(\xi)}|\nabla u(\xi,x)|^2\right)\phi(\xi,x) = \frac{1}{4\epsilon(\xi)} \quad \text{almost surely in } D. \tag{7.4}$$

The discretization of (7.4) uses finite elements for the deterministic dimensions and the gPC for the stochastic dimensions as before. This procedure is analogous to the expositions from Section 6.3. For further details, we refer the reader to [77].

We applied the Ambrosio-Tortorelli segmentation with stochastic parameters on the liver data set from Section 3.4. Again, we initialized the stochastic input image ($n = 1$, $p = 4$) with the expected value of these data. As in the random walker tests, the remaining stochastic dimensions are filled with zeros. To separate the influence of the stochasticity of the parameters in the Ambrosio-Tortorelli model, we use one stochastic parameter for the first test and keep the other parameters deterministic. Figure 7.9 shows the result for a uniformly distributed parameter μ. To be precise, μ is uniformly distributed between 200 and 600, i.e., $\mu \sim \mathcal{U}[200, 600]$. The parameter μ controls the influence of the smoothing term in the image equation. For large μ, we get sharper images with sharp edges. Thus, a stochastic parameter μ influences the smoothing of the image, which is visible from the variance of the smoothed image in Figure 7.9. Here, a smoothing across the object boundaries leads to a variance that looks similar to the original image due to the cartoon-like initial image. Once energy is transported across the edge, it is equally distributed in the whole region due to the smoothing term. The smoothed image influences the phase field because it leads to diffuse boundaries and to a wide phase field that is visible in the phase field variance in Figure 7.9.

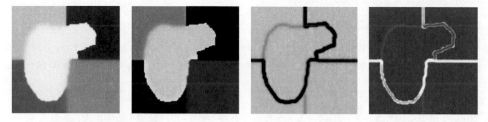

Figure 7.9: Ambrosio-Tortorelli model applied on the expected value of the liver data set using a stochastic parameter μ. *From left to right*: Expected value and variance of the smoothed image and expected value and the variance of the phase field.

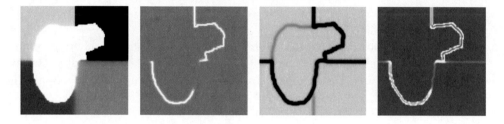

Figure 7.10: Ambrosio-Tortorelli model applied on the expected value of the liver data set using a stochastic parameter ε. *From left to right:* Expected value and variance of the smoothed image and expected value and variance of the phase field.

In Figure 7.10, we used a stochastic parameter ε uniformly distributed between 0.0015 and 0.0035, i.e., $\varepsilon \sim \mathcal{U}[0.0015, 0.0035]$. The parameter ε influences the width of the phase field but has no influence on the smoothing parts of the equations. We observe changes in the variance around the edges. Directly, the parameter ε influences the width of the phase field, and due to the wider phase field, the image is smoothed differently close to the edges.

7.5   GRADIENT-BASED SEGMENTATION

Gradient-based segmentation via a level set formulation contains the parameter ε that controls the influence of the curvature κ. Making this parameter a random variable, we end up with

$$\phi_t(\xi, t, x) + g[u_0](\xi, x)\big(1 - \varepsilon(\xi)\kappa(\xi, t, x)\big)|\nabla\phi(\xi, t, x)| = 0 \quad \text{almost surely in } \mathbb{R}_0^+ \times D$$
$$(7.5)$$

with a stochastic edge indicator function as in (6.36). Again, we use the polynomial chaos to discretize this equation. To deal with the edge indicator and the product $\varepsilon(\xi)\kappa(\xi, t, x)$, we project back onto the polynomial chaos with a Galerkin projection.

We apply the gradient-based segmentation with a stochastic parameter using the CT head data set and the liver data set. As before, we use the expected value as input, one random variable,

$n = 1$, and a polynomial degree $p = 4$. For the experiment, we used a stochastic parameter ε that is uniformly distributed between 0.75 and 1.25, i.e., $\varepsilon \sim \mathcal{U}[0.75, 1.25]$. Since ε controls the influence of the curvature smoothing, higher values for ε lead to smoother contours. This behavior is shown in Figure 7.11 where the contour realizations vary with respect to the curvature.

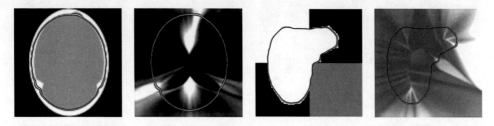

Figure 7.11: Result of the gradient-based segmentation with a stochastic parameter ε, i.e., with a stochastic curvature smoothing. For the CT data set, the expected value of the image along with contour realizations (*left*) and variance of the level set along with contour realizations (*middle left*) are shown. For the liver data set, the same results are shown in the right part of the figure. The red contour corresponds to a $\varepsilon = 0.75$, yellow to $\varepsilon = 1.0$, and blue to $\varepsilon = 1.25$.

7.6   GEODESIC ACTIVE CONTOURS

The sensitivity analysis for the geodesic active contour approach follows the procedure for the sensitivity analysis of the other segmentation methods. For stochastic parameters α, β, and ε, we end up with

$$\phi_t(\xi, t, x) = -\alpha(\xi)\nabla g[u_0](\xi, t, x) \cdot \nabla\phi(\xi, t, x) + \beta(\xi)g[u_0](\xi, t, x)|\nabla\phi(\xi, t, x)|$$
$$+ \varepsilon(\xi)\kappa(\xi, t, x)|\nabla\phi(\xi, t, x)| \quad \text{almost surely in } \mathbb{R}_0^+ \times D.$$

This equation is nearly identical to the stochastic geodesic active contour equation (6.39). However, the numerical treatment with the polynomial chaos requires some additional projection steps back to the polynomial chaos for the products $\alpha\nabla g$, βg, and $\varepsilon\kappa$. Besides these projection steps, we use the same numerical methods as for the discretization of the stochastic geodesic active contour equation in Section 6.4.6, i.e., we use an explicit time-step discretization via the Euler method and a uniform spatial grid.

As an example, we perform the geodesic active contour method with stochastic parameters on the same data sets as in the previous sections. Because we are segmenting smooth objects in the images, we ignore the smoothing term by letting $\varepsilon = 0$. The parabolic approximation and the attraction term $\nabla g \cdot \nabla\phi$ ensure that we also get smooth results with this setting. The parameters α and β shall be uniformly distributed and resulting from the polynomial chaos coefficients $\alpha_1 = 0.08, \alpha_2 = 0.002$, and $\beta_1 = 1.0, \beta_2 = 0.02$. Thus, we use two stochastic parameters simultaneously, and we let them both depend on the same random variable.

Figure 7.12 shows the result for the head CT and the liver data set. The image is easy to segment due to the homogeneous gradient between the inner parts of the head phantom and the bone. The problematic parts are the regions, where the object to segment does not have the "elliptic" contour behavior. In these regions, the gradient differs from the remaining parts of the image. The geodesic active contour method has different attractors depending on the particular value of α and β in these regions. We see this behavior from Figure 7.12, because the contour realizations are far from each other, and the variance is high in a region in the upper part of the object boundary.

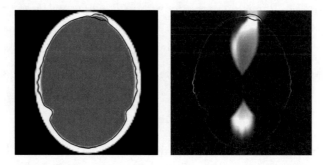

Figure 7.12: Result of the geodesic active contour segmentation with stochastic parameters for the CT data set: expected value of the image and contour realizations (*left*) and the variance of the level set with contour realizations (*right*).

For the liver data set, Figure 7.13 shows the results with the same stochastic parameters. Again, the curves are close together for all parameter realizations, because we have one attractor for the level set only (the object boundary). The differences in the lower part of the object are due to the weak attraction of the liver boundary for some realizations of the parameter α.

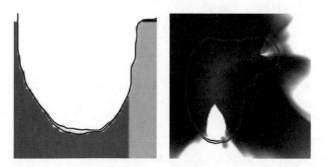

Figure 7.13: Result of the geodesic active contour segmentation with stochastic parameters for the liver data set: Expected value of a detail of the image and contour realizations (*left*) and the variance of the level set with contour realizations (*right*).

Note that for level sets, there is, in contrast to the random walker method, a one-to-one correspondence between the distance between the contour realizations and the variance, because we use a stochastic equivalent of the signed distance function. Thus, deviations in the level set position are directly related to the variance.

7.7   DISCONTINUITY-PRESERVING OPTICAL FLOW

Discontinuity-preserving optical flow can be seen as a combination of the Horn-Schunck optical flow model [44] and the Perona-Malik model [81] for edge- (or discontinuity-) preserving smoothing; see Section 3.6. Referring to the deterministic equations (3.26), we find two parameters α and λ in the edge indicator that control the behavior of the method. Substituting these with random variables $\alpha(\xi)$ and $\lambda(\xi)$ yields the stochastic PDE system

$$\partial_1 u(s,x)\left(\mathbf{w}(\xi,s,x)\cdot\nabla u(s,x)+\partial_s u(s,x)\right)-\alpha(\xi)\mathrm{div}\left(g\left(\xi,|\nabla u_\sigma(s,x)|\right)\nabla w_1(\xi,s,x)\right)=0$$

$$\vdots \qquad\qquad\qquad \vdots$$

$$\partial_d u(s,x)\left(\mathbf{w}(\xi,s,x)\cdot\nabla u(s,x)+\partial_s u(s,x)\right)-\alpha(\xi)\mathrm{div}\left(g\left(\xi,|\nabla u_\sigma(s,x)|\right)\nabla w_d(\xi,s,x)\right)=0$$

almost surely in the space-time cylinder Q. In this equation, the deterministic image sequence $u(s,x)$ is given and the stochastic optical flow field $\mathbf{w}(\xi,s,x):\Gamma\times Q\to\mathbb{R}^d$ is sought. The stochastic edge indicator is exactly as for the Perona-Malik sensitivity analysis, i.e., according to equation (7.3).

We discretize the random parameters by the polynomial chaos as in the previous sections. For the random optical flow field, we utilize the gPC expansion as well. Thus, using test functions $P_j\Psi^\beta$, we end up with a weak form

$$\mathbb{E}\left(\int_D \partial_m u(s,x)\left[\begin{pmatrix}\sum_{\gamma,i}(w_1)_\gamma^i P_i(x)\Psi^\gamma(\xi)\\ \vdots\\ \sum_{\gamma,i}(w_d)_\gamma^i P_i(x)\Psi^\gamma(\xi)\end{pmatrix}\cdot\nabla u(s,x)+\partial_s u(s,x)\right]P_j(x)\Psi^\beta(\xi)\,dx\right.$$

$$\left.+\int_D\sum_{\gamma,\delta,\epsilon,i,k}\left(\alpha_\epsilon\Psi^\epsilon(\xi)\right)\left(g_\delta^k P_k(x)\Psi^\delta(\xi)\right)\left((w_m)_\gamma^i\Psi^\gamma(\xi)\nabla P_i(x)\right)\cdot\left(\Psi^\beta(\xi)\nabla P_j(x)\right)dx\right)=0$$

for every component of $\mathbf{w}$, i.e., $m=1,\ldots,d$, $j\in\mathcal{I}$, $|\beta|=0,\ldots,p$ and for all sequence times $s\in[0,T]$. In this equation, all spatial indices run in $\mathcal{I}$ and all stochastic multi-indices run from absolute value 0 until p.

In the following, we fix one time-point s (i.e., frame) of the sequence time. The matrices associated to the above equation are similar to the ones used above for the stochastic Perona-Malik model in Section 6.1. We define

$$\left(L^{\beta,\gamma}\right)_{i,j}=\sum_k\sum_{\delta,\epsilon}\mathbb{E}\left(\Psi^\beta\Psi^\gamma\Psi^\delta\Psi^\epsilon\right)\alpha_\epsilon g_\delta^k\int_D\nabla P_i(x)\cdot\nabla P_j(x)P_k(x)\,dx,$$

which is the matrix from the stochastic Perona-Malik model with a different weight $\sum_\epsilon \alpha_\epsilon \mathbb{E}\left(\Psi^\beta \Psi^\gamma \Psi^\delta \Psi^\epsilon\right)$. Also, we need the matrices

$$(R_{m,n}^{\beta,\gamma})_{i,j} = \mathbb{E}(\Psi^\beta \Psi^\gamma) \sum_{k,l} u^k u^l \int_D \partial_m P_k(x) P_i(x) \partial_n P_l(x) P_j(x)\, dx$$

as well as the right-hand-side vectors

$$(T_m^\beta)_j = -\mathbb{E}(\Psi^\beta) \sum_{k,l} u^k \tilde{u}^l \int_D \partial_m P_k(x) P_l(x) P_j(x)\, dx,$$

where $\tilde{u}$ serves as an approximation to $\partial_s u$, e.g., by finite differences.

With these matrices at hand, we can write down the full linear system of equations. We denote the vector of unknowns with $(W_1, \ldots, W_d)$, where the component $W_m \in \mathbb{R}^{NM}$ has N stochastic modes in $\mathbb{R}^M$. Also, $T_m \in \mathbb{R}^{NM}$ and so

$$\begin{pmatrix} R_{11} + L & \cdots & R_{1d} \\ \vdots & & \vdots \\ R_{d1} & \cdots & R_{dd} + L \end{pmatrix} \begin{pmatrix} W_1 \\ \vdots \\ W_d \end{pmatrix} = \begin{pmatrix} T_1 \\ \vdots \\ T_d \end{pmatrix},$$

where the blocks $R_{mn} = (R_{mn}^{\beta,\gamma})_{|\beta|,|\gamma|=0,\ldots,p}$ and $L = (L^{\beta,\gamma})_{|\beta|,|\gamma|=0,\ldots,p}$ are composed of $N \times N$ blocks of size $\mathbb{R}^{M \times M}$.

As a demonstration, we perform the sensitivity analysis for discontinuity-preserving optical flow on two data sets with ten frames each. The first data set shows a circle moving to the top in a 100×100 pixel image. We use a uniformly distributed smoothing parameter, $\alpha \sim \mathcal{U}[16, 24]$ ($\mathbb{E}(\alpha) = 20$, coefficients $\alpha_1 = 20$, $\alpha_2 = 4$), and a uniformly distributed parameter $\lambda \sim \mathcal{U}[0.08, 0.12]$, for Perona-Malik smoothing (coefficients $\lambda_1 = 0.1$ and $\lambda_2 = 0.02$). The results of this computation are depicted in Figure 7.14. Clearly, there is a high optical flow on the object boundaries. Due to the smoothing and the homogeneous background, we get a big halo in the flow field around the object. The random variable α describes the intensity of this global smoothing, and the random variable λ influences the smoothing process on the object boundaries only. This effect is also visible in the covariance $|\mathrm{Cov}(w)|_\infty = \max(\mathrm{Var}\, u, \mathrm{Var}\, v)$ of the optical flow.

Furthermore, we apply the sensitivity analysis for optical flow estimation on a test sequence with resolution 200×200 pixels from [67], using the smoothing parameters from above. The first and last images of the test sequence, the expected value and covariance of the optical flow field, are depicted in Figure 7.15. The expected value of the flow field shows that the discontinuity-preserving optical flow model gives a rough expression of the optical flow in the image. Using the covariance, we are able to identify regions where the smoothing term has a significant influence on the optical flow estimation. The regions with a high covariance of the optical flow field are the smooth regions in the image, because there is no edge information that gives additional "source-terms" for the optical flow.

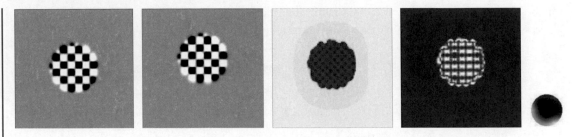

Figure 7.14: The first and last frames of the input sequence (*left and middle left*), expected value (*middle right*), and maximal covariance (*right*) of the optical flow field. The colorwheel at the *far right* indicates the directions of the flow field.

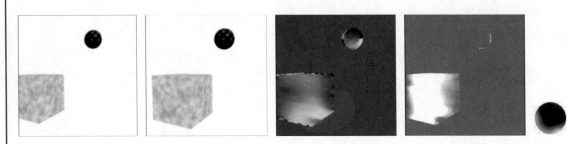

Figure 7.15: The first and last frames of the input sequence (*left and middle left*), expected value (*middle right*), and maximal covariance (*right*) of the optical flow field. The colorwheel at the *far right* indicates the directions of the flow field.

Note 7.1   Readers interested in more information about sensitivity analysis with the presented framework are referred to this literature:

- [2]: Sensitivity analysis of a coupled nonlinear PDE system with a medical application

- [78]: Details about several applications presented within this chapter

CHAPTER 8

Conclusions

The goal of this book was to introduce the reader to the recent advances in the field of uncertainty quantification and error propagation for computer vision, image processing, and image analysis that are based on partial differential equations (PDEs). Our original motivation was the observation that more and more often, measured data in the form of images are being used as part of the simulation pipeline, and that error propagation and uncertainty quantification—starting from images but then moving through the pipeline—are important to understanding the ending result of this scientific or engineering process. To analyze the image processing pipeline, we have presented concepts that enable error propagation to be formulated with a set of basic operations: the idea of stochastic images and corresponding (numerical PDE) operations on those images. From this perspective, one approach commonly used within the engineering literature is to employ Monte Carlo sampling techniques to attempt to quantify the impact of errors and variability on an engineering pipeline. Within the uncertainty quantification world, alternative approaches such as the class of generalized polynomial chaos (gPC) methods, which make assumptions about the ability to approximate the process of interest through polynomial approximations, have gained traction as a way of accelerating the convergence of the error and uncertainty quantification process in a computationally tractable way. In this book, we have shown the results of our exploration of the use of the gPC methodology for image processing and computer vision problems. We have relied on the fact that all the prerequisites of the gPC framework (e.g., finite variance, smoothness in the stochastic space, etc.) are intrinsically satisfied in the image processing context. In particular, we have shown how the generalized polynomial chaos (gPC) approach and its corresponding rules for computation (e.g., sums, products, projections on the gPC space, etc.) lead to straightforward generalizations to image processing and computer vision techniques. We have shown that the gPC methodology, when combined with image processing and computer vision techniques, provides a powerful approach to error propagation and uncertainty quantification, and to accessing system sensitivity.

We have focused on intrusive algorithms, and we briefly touched the space of nonintrusive (sample-based) algorithms for uncertainty quantification of stochastic images only. There is certainly room for further investigation here. We also see the need for better visualization tools that enable exploration of the types of data these uncertainty quantifications generate. Furthermore, we see the need to bring these methods into practice: that is, to apply the pipeline described herein to "clinical" applications—those used day-to-day in laboratories and clinics that regularly employ imaging sciences as part of workflow. We emphasize, once more, that all that has been presented in this book has no limitation with respect to practical importance in all fields of research and application in which digital images are being used to yield quantitative information.

Bibliography

[1] R. P. Adams and D. J. MacKay. *Bayesian Online Changepoint Detection*. Cambridge, UK, 2007.

[2] I. Altrogge, T. Preusser, T. Kröger, S. Haase, T. Pätz, and R. M. Kirby. Sensitivity analysis for the optimization of radiofrequency ablation in the presence of material parameter uncertainty. *International Journal for Uncertainty Quantification*, 2(3):295–321, 2012. DOI: 10.1615/int.j.uncertaintyquantification.2012004135.

[3] L. Alvarez, F. Guichard, P. L. Lions, and J. M. Morel. Axioms and fundamental equations of image processing. *Archive for Rational Mechanics and Analysis*, 123:199–257, 1993. DOI: 10.1007/bf00375127.

[4] L. Ambrosio and M. Tortorelli. Approximation of functionals depending on jumps by elliptic functionals via Gamma-convergence. *Communications on Pure and Applied Mathematics*, 43(8):999–1036, 1990. DOI: 10.1002/cpa.3160430805.

[5] R. Askey and J. Wilson. Some basic hypergeometric polynomials that generalize Jacobi polynomials. *Memoirs of the American Mathematical Society*, 319, 1985.

[6] G. Aubert and P. Kornprobst. *Mathematical Problems in Image Processing: Partial Differential Equations and the Calculus of Variations*, 2nd ed., volume 147 of *Applied Mathematical Sciences*. Springer-Verlag, 2006.

[7] I. Babuška, R. Tempone, and G. E. Zouraris. Solving elliptic boundary value problems with uncertain coefficients by the finite element method: The stochastic formulation. *Computer Methods in Applied Mechanics and Engineering*, 194(12-16):1251–1294, 2005. Special Issue on Computational Methods in Stochastic Mechanics and Reliability Analysis. DOI: 10.1016/j.cma.2004.02.026.

[8] J. Beutel. *Handbook of Medical Imaging: Physics and Psychophysics*. SPIE Press, 2000. DOI: 10.1117/3.832716.

[9] B. Blackmore. Validation and sensitivity analysis of an image processing technique to derive thermal conductivity variation within a printed circuit board. In *25th Semiconductor Thermal Measurement and Management Symposium*, pages 76–86, 2009. DOI: 10.1109/stherm.2009.4810746.

[10] A. Bruhn, J.Weickert, and C.Schnörr. Lucas/kanade meets horn/schunck: Combining local and global optic flow methods. *International Journal of Computer Vision*, 61(3), 2005. DOI: 10.1023/b:visi.0000045324.43199.43.

[11] D. Cacuci, M. Ionescu-Bujor, and I. Navon. *Sensitivity and Uncertainty Analysis: Applications to Large-scale Systems*. Chapman & Hall/CRC Press, 2005. DOI: 10.1201/9780203483572.

[12] R. H. Cameron and W. T. Martin. The orthogonal development of non-linear functionals in series of Fourier-Hermite functionals. *The Annals of Mathematics*, 48(2):385–392, 1947. DOI: 10.2307/1969178.

[13] E. Carlstein, H. Müller, and D. Siegmund. Change-point problems. In *Change-point Problems*, number 23. Institute of Mathematical Statistics, 1994.

[14] V. Caselles, F. Catte, B. Coll, and F. Dibos. A geometric model for edge detection. *Numerische Mathematik*, 66:1–31, 1993.

[15] V. Caselles, R. Kimmel, and G. Sapiro. Geodesic active contours. *International Journal of Computer Vision*, 22(1):61–79, 1997. DOI: 10.1109/iccv.1995.466871.

[16] F. Catté, P.-L. Lions, J.-M. Morel, and T. Coll. Image selective smoothing and edge detection by nonlinear diffusion. *SIAM Journal of Numerical Analysis*, 29(1):182–193, 1992. DOI: 10.1137/0729012.

[17] T. Chan and J. Shen. *Image Processing and Analysis: Variational, PDE, Wavelet, and Stochastic Methods*. Society for Industrial and Applied Mathematics, Philadelphia, PA, U.S., 2005.

[18] T. Chan and L. Vese. Active contours without edges. *IEEE Transactions on Image Processing*, 10(2):266–277, 2001. DOI: 10.1109/83.902291.

[19] C. W. Clenshaw and A. R. Curtis. A method for numerical integration on an automatic computer. *Numerische Mathematik*, 2:197–205, 1960. DOI: 10.1007/bf01386223.

[20] B. J. Debusschere, H. N. Najm, P. P. Pébay, O. M. Knio, R. G. Ghanem, and O. P. Le Maître. Numerical challenges in the use of polynomial chaos representations for stochastic processes. *SIAM Journal on Scientific Computing*, 26(2):698–719, 2005. DOI: 10.1137/s1064827503427741.

[21] A. Dervieux and F. Thomasset. A finite element method for the simulation of a Rayleigh-Taylor instability. In R. Rautmann, Ed., *Approximation Methods for Navier-Stokes Problems*, volume 771 of *Lecture Notes in Mathematics*, pages 145–158. Springer, 1980. DOI: 10.1007/bfb0086897.

[22] C. Desceliers, R. Ghanem, and C. Soize. Maximum likelihood estimation of stochastic chaos representations from experimental data. *International Journal for Numerical Methods in Engineering*, 66:978–1001, 2006. DOI: 10.1002/nme.1576.

[23] S. Djurcilov, K. Kim, P. F. J. Lermusiaux, and A. Pang. Volume rendering data with uncertainty information. *Data Visualization (Proc. of the EG+IEEE VisSym)*, pages 243–252, 2001. DOI: 10.1007/978-3-7091-6215-6_26.

[24] B. Echebarria, R. Folch, A. Karma, and M. Plapp. Quantitative phase-field model of alloy solidification. *Physical Review E*, 70(6):061604, 2004. DOI: 10.1103/physreve.70.061604.

[25] E. Eisenhauer, P. Therasse, J. Bogaerts, L. Schwartz, D. Sargent, R. Ford, J. Dancey, S. Arbuck, S. Gwyther, M. Mooney, L. Rubinstein, L. Shankar, L. Dodd, R. Kaplan, D. Lacombe, and J. Verweij. New response evaluation criteria in solid tumours: Revised recist guideline (version 1.1). *European Journal of Cancer*, 45(2):228–247, 2009. DOI: 10.1016/j.ejca.2008.10.026.

[26] E. Erdem, A. Sancar-Yilmaz, and S. Tari. Mumford-Shah regularizer with spatial coherence. In F. Sgallari, A. Murli, and N. Paragios, Eds., *Scale Space and Variational Methods in Computer Vision*, volume 4485 of *Lecture Notes in Computer Science*, pages 545–555. Springer Berlin/Heidelberg, 2007. DOI: 10.1007/978-3-540-72823-8.

[27] O. G. Ernst, A. Mugler, H.-J. Starkloff, and E. Ullmann. On the convergence of generalized polynomial chaos expansions. *ESAIM: Mathematical Modelling and Numerical Analysis*, 46:317–339, 2012. DOI: 10.1051/m2an/2011045.

[28] R. Folch, J. Casademunt, A. Hernandez-Machado, and L. Ramirez-Piscina. Phase-field model for Hele-Shaw flows with arbitrary viscosity contrast. I. Theoretical approach. *Physical Review E*, 60:1724, 1999. DOI: 10.1103/physreve.60.1724.

[29] K. Frank and S. Heinrich. Computing discrepancies of Smolyak quadrature rules. *Journal of Complexity*, 12(4):287–314, 1996. DOI: 10.1006/jcom.1996.0020.

[30] D. Gabor. Information theory in electron microscopy. *Laboratory Investigation*, 14:801–807, 1965.

[31] D. Gilbarg and N. Trudinger. *Elliptic Partial Differential Equations of Second Order*. Classics in Mathematics. U.S. Government Printing Office, 2001. DOI: 10.1007/978-3-642-61798-0.

[32] J.-F. Giovanelli and J. Idier, Eds. *Regularization and Bayesian Methods for Inverse Problems in Signal and Image Processing*. Wiley, 2015. DOI: 10.1002/9781118827253.

[33] K. Glasner. Nonlinear preconditioning for diffuse interfaces. *Journal of Computational Physics*, 174(2):695–711, 2001. DOI: 10.1006/jcph.2001.6933.

[34] G. H. Golub and C. F. V. Loan. *Matrix Computations*, 3rd ed. The Johns Hopkins University Press, 1996.

[35] L. Grady. Multilabel randomwalker image segmentation using prior models. In *Proc. of CVPR*, volume 1, pages 763–770, 2005. DOI: 10.1109/cvpr.2005.239.

[36] L. Grady. Random walks for image segmentation. *IEEE Transactions on Pattern Analysis and Machine Intelligence*, 28(11):1768–1783, 2006. DOI: 10.1109/tpami.2006.233.

[37] H. Griethe and H. Schumann. The visualization of uncertain data: Methods and problems. In T. Schulze, G. Horton, B. Preim, and S. Schlechtweg, Eds., *Simulation und Visualisierung 2006 (SimVis 2006), Magdeburg*, pages 143–156. SCS Publishing House e.V., 2006.

[38] H. Gudbjartsson and S. Patz. The rician distribution of noisy MRI data. *Magnetic Resonance in Medicine*, 34:910?914, 1995. DOI: 10.1002/mrm.1910340618.

[39] H. Harbrecht, M. Peters, and R. Schneider. On the low-rank approximation by the pivoted Cholesky decomposition. *Applied Numerical Mathematics*, 62(4):428–440, 2012. DOI: 10.1016/j.apnum.2011.10.001.

[40] G. Herman. *Fundamentals of Computerized Tomography: Image Reconstruction from Projections*. Advances in pattern recognition. Springer, 2009. DOI: 10.1007/978-1-84628-723-7.

[41] M. R. Hestenes and E. Stiefel. Methods of conjugate gradients for solving linear systems. *Journal of Research of the National Bureau of Standards*, 49(6):409–436, 1952. DOI: 10.6028/jres.049.044.

[42] T. Hida and N. Ikeda. Analysis on Hilbert space with reproducing kernel arising from multiple Wiener integral. In L. M. Le Cam and J. Neyman, Eds., *Proc. of the 5th Berkeley Symposium on Mathematical Statistics and Probability*. University of California Press, 1967.

[43] H. Holden, B. Øksendal, J. Ubøe, and T. Zhang. *Stochastic Partial Differential Equations. A Modeling, White Noise Functional Approach*, 2nd ed. Universitext. New York, Springer, 2010. DOI: 10.1007/978-1-4684-9215-6_4.

[44] B. K. P. Horn and B. G. Schunck. Determining optical flow. *Artificial Intelligence*, 17:185–203, 1981. DOI: 10.1016/0004-3702(81)90024-2.

[45] T. Iijima. Basic theory of pattern observation. *Papers of Technical Group on Automata and Automatic Control, IECE*, 1959.

[46] I. James. *The Topology of Stiefel Manifolds*. London Mathematical Society lecture note series. Cambridge University Press, 1976. DOI: 10.1017/cbo9780511600753.

[47] S. Janson. *Gaussian Hilbert Spaces*. Cambridge University Press, 1997. DOI: 10.1017/cbo9780511526169.

[48] C. Johnson. *Numerical Solution of Partial Differential Equations by the Finite Element Method*. Dover Books on Mathematics. Dover Publications, 2009.

[49] I. Jolliffe. *Principal Component Analysis*. Springer, 2002. DOI: 10.1007/978-1-4757-1904-8.

[50] G. Karniadakis and R. Kirby. *Parallel Scientific Computing in C++ and MPI: A Seamless Approach to Parallel Algorithms and their Implementation*. Cambridge University Press, 2003. DOI: 10.1017/cbo9780511812583.

[51] M. Kass, A. Witkin, and D. Terzopoulos. Snakes: Active contour models. *International Journal of Computer Vision*, 1(4):321–331, 1988. DOI: 10.1007/bf00133570.

[52] M. Kendall, A. Stuart, J. Ord, and A. O'Hagan. *Kendall's Advanced Theory of Statistics*, volume 1. Edward Arnold, 1994.

[53] S. Kichenassamy, A. Kumar, P. Olver, A. Tannenbaum, and A. Yezzi. Gradient flows and geometric active contour models. In *Proc. of the 5th International Conference on Computer Vision*, pages 810–815, 1995. DOI: 10.1109/iccv.1995.466855.

[54] J. J. Koenderink. The structure of images. *Biological Cybernetics*, 50:363–370, 1984. DOI: 10.1007/bf00336961.

[55] Y. G. Kondratiev, P. Leukert, and L. Streit. Wick calculus in Gaussian analysis. *Acta Applicandae Mathematicae*, 44:269–294, 1996.

[56] D. Landau and K. Binder. *A Guide to Monte Carlo Simulations in Statistical Physics*. Cambridge University Press, 2005. DOI: 10.1017/cbo9780511614460.

[57] C. Li, C. Xu, C. Gui, and M. D. Fox. Level set evolution without re-initialization: A new variational formulation. In *IEEE Computer Society Conference on Computer Vision and Pattern Recognition (CVPR)*, pages 430–436, 20–26, June 2005. DOI: 10.1109/cvpr.2005.213.

[58] Y. Lin, B. Wohlberg, and H. Guo. UPRE method for total variation parameter selection. *Signal Processing*, 90(8):2546–2551, 2010. DOI: 10.1016/j.sigpro.2010.02.025.

[59] M. Ljungberg, S. Strand, and M. King. *Monte Carlo Calculations in Nuclear Medicine: Applications in Diagnostic Imaging*. Medical Science Series. Taylor & Francis, 1998. DOI: 10.1201/b13073.

[60] M. Loève. *Probability Theory*, 4th ed. Springer-Verlag, New York, 1977. DOI: 10.1007/978-1-4684-9464-8.

[61] B. Lucas and T. Kanade. An iterative image registration technique with an application to stereo vision. In *DARPA Image Understanding Workshop*, pages 121–130, 1981.

[62] O. Maitre and O. Knio. *Spectral Methods for Uncertainty Quantification: With Applications to Computational Fluid Dynamics*. Scientific Computation. Springer Netherlands, 2010. DOI: 10.1007/978-90-481-3520-2.

[63] R. Malladi, J. A. Sethian, and B. C. Vemuri. Evolutionary fronts for topology-independent shape modeling and recovery. In *Proc. of the 3rd European Conference on Computer Vision (ECCV'94)*, volume 1, pages 3–13. Springer-Verlag New York, 1994. DOI: 10.1007/3-540-57956-7_1.

[64] M. Matsumoto and T. Nishimura. Mersenne twister: A 623-dimensionally equidistributed uniform pseudo-random number generator. *ACM Transactions on Modeling and Computer Simulation*, 8(1):3–30, 1998. DOI: 10.1145/272991.272995.

[65] H. G. Matthies and B. V. Rosic. Inelastic media under uncertainty: Stochastic models and computational approaches. In B. D. Reddy, Ed., *IUTAM Symposium on Theoretical, Computational and Modelling Aspects of Inelastic Media*, volume 11 of *IUTAM Bookseries*, pages 185–194. Springer Netherlands, 2008. DOI: 10.1007/978-1-4020-9090-5.

[66] B. McCane, K. Novins, D. Crannitch, and B. Galvin. On benchmarking optical flow. *Computer Vision and Image Understanding*, 84(1):126–143, 2001. DOI: 10.1006/cviu.2001.0930.

[67] B. McCane, K. Novins, D. Crannitch, and B. Galvin. On benchmarking optical flow. *Computer Vision and Image Understanding*, 84(1):126–143, 2001. DOI: 10.1006/cviu.2001.0930.

[68] J. Modersitzki. *Numerical Methods for Image Registration*. Numerical mathematics and scientific computation. Oxford University Press, 2004. DOI: 10.1093/acprof:oso/9780198528418.001.0001.

[69] J. Modersitzki. *FAIR: Flexible Algorithms for Image Registration*. Fundamentals of Algorithms. Society for Industrial and Applied Mathematics (SIAM, 3600 Market Street, Floor 6, Philadelphia, PA 19104), 2009. DOI: 10.1137/1.9780898718843.

[70] D. Mumford and J. Shah. Optimal approximations by piecewise smooth functions and associated variational problems. *Communications on Pure and Applied Mathematics*, 42(5):577–685, 1989. DOI: 10.1002/cpa.3160420503.

[71] M. Musiela and T. Zariphopoulou. Stochastic partial differential equations and portfolio choice. In *Contemporary Quantitative Finance*, pages 195–216. Springer, 2010. DOI: 10.1007/978-3-642-03479-4_11.

[72] A. Nouy. A generalized spectral decomposition technique to solve a class of linear stochastic partial differential equations. *Computer Methods in Applied Mechanics and Engineering*, 196(45-48):4521–4537, 2007. DOI: 10.1016/j.cma.2007.05.016.

[73] E. Novak and K. Ritter. The curse of dimension and a universal method for numerical integration. In *Multivariate Approximation and Splines*, pages 177–187, 1998. DOI: 10.1007/978-3-0348-8871-4_15.

[74] J. E. Oakley and A. O'Hagan. Probabilistic sensitivity analysis of complex models: A Bayesian approach. *Journal of the Royal Statistical Society, Series B*, 66:751–769, 2002. DOI: 10.1111/j.1467-9868.2004.05304.x.

[75] S. Osher and J. A. Sethian. Fronts propagating with curvature dependent speed: Algorithms based on Hamilton-Jacobi formulations. *Journal of Computational Physics*, 79:12–49, 1988. DOI: 10.1016/0021-9991(88)90002-2.

[76] T. Pätz, R. M. Kirby, and T. Preusser. Ambrosio-Tortorelli segmentation of stochastic images: Model extensions, theoretical investigations and numerical methods. *International Journal of Computer Vision*, 103(2), 2013. DOI: 10.1007/s11263-012-0578-8.

[77] T. Pätz and T. Preusser. Ambrosio-Tortorelli segmentation of stochastic images. In K. Daniilidis, P. Maragos, and N. Paragios, Eds., *Computer Vision (ECCV)*, volume 6315 of *Lecture Notes in Computer Science*, pages 254–267. Springer Berlin/Heidelberg, 2010. (This paper received the *ECCV 2010 Best Student Paper Award*.) DOI: 10.1007/978-3-642-15558-1.

[78] T. Pätz and T. Preusser. Fast parameter sensitivity analysis of PDE-based image processing methods. In *ECCV*, volume 7578 of *Lecture Notes in Computer Science*, pages 140–153, 2012. DOI: 10.1007/978-3-642-33786-4_11.

[79] T. Pätz and T. Preusser. Segmentation of stochastic images with a stochastic random walker method. *IEEE Transactions on Image Processing*, 21(5):2424–2433, 2012. DOI: 10.1109/tip.2012.2187531.

[80] P. Perona and J. Malik. Scale space and edge detection using anisotropic diffusion. *IEEE Transactions Pattern Analysis Machine Intelligence*, 12:629–639, 1990. DOI: 10.1109/34.56205.

[81] P. Perona and J. Malik. Scale-space and edge detection using anisotropic diffusion. *IEEE Transactions on Pattern Analysis and Machine Intelligence*, 12(7):629–639, July 1990. DOI: 10.1109/34.56205.

[82] K. Petras. *Asymptotically Minimal Smolyak Cubature*. Technical report, Technische Universität Braunschweig, 1999.

[83] H. Pirsiavash, S. Kasaei, and F. Marvasti. An efficient parameter selection criterion for image denoising. In *Proc. of the 5th IEEE ISSPIT*, pages 872–877, Dec. 2005. DOI: 10.1109/isspit.2005.1577214.

[84] K. Potter, J. Krüger, and C. Johnson. Towards the visualization of multi-dimensional stochastic distribution data. In *Proc. of the International Conference on Computer Graphics and Visualization (IADIS)*, 2008.

[85] J.-S. Prassni, T. Ropinski, and K. Hinrichs. Uncertainty-aware guided volume segmentation. *IEEE Transactions on Visualization and Computer Graphics*, 16:1358–1365, 2010. DOI: 10.1109/tvcg.2010.208.

[86] T. Preusser and M. Rumpf. An adaptive finite element method for large scale image processing. *Journal of Visual Communication and Image Representation*, 11(2):183–195, 2000. DOI: 10.1006/jvci.1999.0444.

[87] A. Quarteroni, R. Sacco, and F. Saleri. *Numerical Mathematics*. Texts in applied mathematics. Springer, 2000. DOI: 10.1007/b98885.

[88] S. Rajan, S. Wang, R. Inkol, and A. Joyal. Efficient approximations for the arctangent function. *IEEE Signal Processing Magazine*, 23(3):108–111, 2006. DOI: 10.1002/9780470170090.ch18.

[89] D. W. O. Rogers. Fifty years of Monte Carlo simulations for medical physics. *Physics in Medicine and Biology*, 51(13):R287–R301, 2006. DOI: 10.1088/0031-9155/51/13/r17.

[90] K. M. Rosenberg. *CTSim—Open Source Computed Tomography Simulator*. http://ctsim.org

[91] A. Saltelli, K. Chan, and E. Scott. *Sensitivity Analysis*. Wiley series in probability and statistics. Wiley, 2000. DOI: 10.1007/978-3-642-04898-2_509.

[92] G. Sapiro. *Geometric Partial Differential Equations and Image Analysis*. Cambridge University Press, New York, 2006. DOI: 10.1017/cbo9780511626319.

[93] O. Scherzer, M. Grasmair, H. Grossauer, M. Haltmeier, and F. Lenzen. *Variational Methods in Imaging*. Springer, 2009.

[94] J. A. Sethian. *Level Set Methods and Fast Marching Methods*. Cambridge University Press, 1999.

[95] L. Shepp and B. Logan. The Fourier reconstruction of a head section. *IEEE Transactions on Nuclear Science*, 21(3):21–43, 1974. DOI: 10.1109/tns.1974.6499235.

[96] S. Smolyak. Quadrature and interpolation formulas for tensor products of certain classes of functions. *Soviet Mathematics—Doklady*, 4:240–243, 1963.

[97] G. Stefanou, A. Nouy, and A. Clement. Identification of random shapes from images through polynomial chaos expansion of random level-set functions. *International Journal for Numerical Methods in Engineering*, 79(2):127–155, 2009. DOI: 10.1002/nme.2546.

[98] T. R. Steger, R. A. White, and E. F. Jackson. Input parameter sensitivity analysis and comparison of quantification models for continuous arterial spin labeling. *Magnetic Resonance in Medicine*, 53(4):895–903, 2005. DOI: 10.1002/mrm.20440.

[99] Y. Sun and C. Beckermann. Sharp interface tracking using the phase-field equation. *Journal of Computational Physics*, 220(2):626–653, 2007. DOI: 10.1016/j.jcp.2006.05.025.

[100] P. Therasse, S. G. Arbuck, E. A. Eisenhauer, J. Wanders, R. S. Kaplan, L. Rubinstein, J. Verweij, M. Van Glabbeke, A. T. van Oosterom, M. C. Christian, and S. G. Gwyther. New guidelines to evaluate the response to treatment in solid tumors. *Journal of the National Cancer Institute*, 92(3):205–216, 2000. DOI: 10.1093/jnci/92.3.205.

[101] H. Tiesler, R. M. Kirby, D. Xiu, and T. Preusser. Stochastic collocation for optimal control problems with stochastic PDE constraints. *Submitted to SIAM Journal on Optimization*, 2010. DOI: 10.1137/110835438.

[102] R. Tsai and S. Osher. Level set methods and their applications in image science. *Communications in Mathematical Sciences*, 1(4):623–656, 2003. DOI: 10.4310/cms.2003.v1.n4.a1.

[103] G. Turk and M. Levoy. Zippered polygon meshes from range images. In *Proc. of the 21st Annual Conference on Computer Graphics and Interactive Techniques, (SIGGRAPH)*, pages 311–318, 1994. DOI: 10.1145/192161.192241.

[104] D. B. Twieg. The k-trajectory formulation of the NMR imaging process with applications in analysis and synthesis of imaging methods. *Medical Physics*, 10(5):610–621, 1983. DOI: 10.1118/1.595331.

[105] G. Vage. Variational methods for PDEs applied to stochastic partial differential equations. *Mathematica Scandinavica*, 82:113–137, 1998. DOI: 10.7146/math.scand.a-13828.

[106] P. Viola and W. M. Wells III. Alignment by maximization of mutual information. *International Journal of Computer Vision*, 24(2):137–154, 1997. DOI: 10.1109/iccv.1995.466930.

[107] G. G. Walter. *Wavelets and Other Orthogonal Systems with Applications*. CRC Press, 1994.

[108] J. Weickert. *Anisotropic Diffusion in Image Processing*. B. G. Teubner, Stuttgart, 1998.

[109] M. Wernick and J. Aarsvold. *Emission Tomography: The Fundamentals of PET and SPECT*. Elsevier Academic Press, 2004.

[110] G. C. Wick. The evaluation of the collision matrix. *Physical Review*, 80(2):268–272, 1950. DOI: 10.1103/physrev.80.268.

[111] N. Wiener. The homogeneous chaos. *American Journal of Mathematics*, 60(4):897–936, 1938. DOI: 10.2307/2371268.

[112] A. P. Witkin. Scale-space filtering. In *Proc. of the International Joint Conference on Artificial Intelligence*, pages 1019–1021. ACM Inc. New York, 1983. DOI: 10.1016/b978-0-08-051581-6.50036-2.

[113] C. S. Won and R. M. Gray. *Stochastic Image Processing*. Kluver Academic, 2004. DOI: 10.1007/978-1-4419-8857-7.

[114] D. Xiu. Fast numerical methods for stochastic computations: A review. *Communications in Computational Physics*, 5(2–4):242–272, 2009.

[115] D. Xiu. *Numerical Methods for Stochastic Computations: A Spectral Method Approach*. Princeton University Press, 2010.

[116] D. Xiu and G. E. Karniadakis. The Wiener–Askey polynomial chaos for stochastic differential equations. *SIAM Journal on Scientific Computing*, 24(2):619–644, 2002. DOI: 10.1137/s1064827501387826.

[117] Y. Zhang, D. B. Goldgof, S. Sarkar, and L. V. Tsap. A sensitivity analysis method and its application in physics-based nonrigid motion modeling. *Image and Vision Computing*, 25(3):262–273, 2007. DOI: 10.1016/j.imavis.2005.08.007.

[118] X. Zhu and P. Milanfar. Automatic parameter selection for denoising algorithms using a no-reference measure of image content. *IEEE T Image Processing*, 19(12):3116–3132, 2010. DOI: 10.1109/tip.2010.2052820.

[119] H. Zimmer, A. Bruhn, and J. Weickert. Optic flow in harmony. *International Journal of Computer Vision*, 93(3):368–388, 2011. DOI: 10.1007/s11263-011-0422-6.

Authors' Biographies

TOBIAS PREUSSER

Tobias Preusser studied mathematics at the University of Bonn and at New York University. He received his Ph.D. from the University of Duisburg with a thesis on anisotropic geometric diffusion in image processing and his habilitation in Mathematics from the University of Bremen with a thesis on image-based computing. He is a professor for the mathematical modeling of biomedical processes at Jacobs University Bremen, head of the modeling and simulation group, and member of the management board at the Fraunhofer Institute for Medical Image Computing MEVIS. On the one hand, his research interests include modeling and simulation of bio-medical processes with partial differential equations (PDEs), mathematical image processing, and scientific visualization based on PDEs. On the other hand, his research is driven by concrete and complex application problems in medicine.

ROBERT M. KIRBY

Robert M. (Mike) Kirby received an M.S. degree in applied mathematics, an M.S. degree in computer science, and a Ph.D. degree in applied mathematics from Brown University, Providence, Rhode Island, in 1999, 2001, and 2002, respectively. He is currently a Professor of Computing and Associate Director of the School of Computing, University of Utah, Salt Lake City, where he is also an Adjunct Professor in the Departments of Bioengineering and Mathematics and a member of the Scientific Computing and Imaging Institute. His current research interests include scientific computing and visualization.

TORBEN PÄTZ

Torben Pätz received his diploma degree in mathematics from the University of Bremen, Germany, in 2009 with a thesis focusing on the numerical simulation of radio-frequency ablation and his Ph.D. from Jacobs University Bremen, Germany, in 2009. In his Ph.D. thesis, he focused on the segmentation of stochastic images with stochastic PDEs, laying the basis for the book at hand. He was a postdoctoral fellow at Jacobs University Bremen while writing the book and is now a research scientist at the Fraunhofer Institute for Medical Image Computing MEVIS. His research interests include uncertainty modeling and propagation in medical applications as well as software support for interventional radiology.

Printed in the United States
by Baker & Taylor Publisher Services